MENTAL COMPUTATION AND ESTIMATION

Mental calculations and estimations are basic, everyday skills that are essential for real-life arithmetic operations and number sense. This book presents a much needed overview and analysis of mental computation and estimation, drawing on contemporary research and empirical studies that were conducted on students, teachers and adults to cover all aspects of this complex field.

Mental Computation and Estimation analyses the implications of the research, teaching and learning of mathematics and delivers effective practices that will enhance everyday learning for students. Focusing on a range of international research and studies from the School of Nature and Life Mathematics in Greece, it answers a number of important questions including:

- What are mental calculations and estimations, why they are important and what other mathematical concepts and cognitive behaviours are they related to?
- What strategies are used in mental addition, subtraction, multiplication and division and how are multiplication tables learned?
- What are the new trends in the teaching of mental calculation and estimation?

An invaluable resource for all those involved in the practice and research of mathematics education, *Mental Computation and Estimation* will also be a useful tool for researchers, policy makers and developers of educational programs.

Charalampos Lemonidis is Professor of Mathematics Education. He is currently Director of Postgraduate Studies in the Department of Primary Education, University of Western Macedonia, Greece, and has founded the School of Nature and Life Mathematics.

to Polymnia

MENTAL COMPUTATION AND ESTIMATION

Implications for mathematics education research, teaching and learning

Charalampos Lemonidis

LONDON AND NEW YORK

First published 2016
by Routledge
2 Park Square, Milton Park, Abingdon, Oxon OX14 4RN

and by Routledge
711 Third Avenue, New York, NY 10017

Routledge is an imprint of the Taylor & Francis Group, an informa business

© 2016 Charalampos Lemonidis

The right of Charalampos Lemonidis to be identified as author of this work has been asserted by him in accordance with sections 77 and 78 of the Copyright, Designs and Patents Act 1988.

All rights reserved. No part of this book may be reprinted or reproduced or utilised in any form or by any electronic, mechanical, or other means, now known or hereafter invented, including photocopying and recording, or in any information storage or retrieval system, without permission in writing from the publishers.

Trademark notice: Product or corporate names may be trademarks or registered trademarks, and are used only for identification and explanation without intent to infringe.

British Library Cataloguing in Publication Data
A catalogue record for this book is available from the British Library

Library of Congress Cataloging in Publication Data
Lemonidis, Charalampos.
Mental computation and estimation: implications for mathematics education research, teaching and learning
Charalampos Lemonidis.
pages cm
1. Mental arithmetic. 2. Arithmetic—Study and teaching. 3. Reasoning. I. Title.
QA109.L46 2016
513'.9—dc23
2015021806

ISBN: 978-1-138-93842-7 (hbk)
ISBN: 978-1-138-93843-4 (pbk)
ISBN: 978-1-315-67566-4 (ebk)

Typeset in Bembo
by Swales & Willis Ltd, Exeter, Devon, UK

CONTENTS

Acknowledgements		*vi*
	Introduction	1
1	Theoretical issues concerning mental calculation	6
2	Teaching mental calculation	44
3	Mental calculation: addition and subtraction	73
4	Mental calculation: multiplication and division	115
5	Mental calculation with rational numbers	159
6	Computational estimation	183
7	Learning difficulties and mental calculation	207
Index		*233*

ACKNOWLEDGEMENTS

I am very grateful to all my postgraduate students as well as to Alexandra Anastasiadou, Anastasia Masaraki, Kyriaki Thoidou and Dora Xostelidou for their assistance in the translation of the original text into English.

INTRODUCTION

In contemporary education, students are best served by doing useful mathematics they can understand, rather than mathematics that they do not understand and treat in a mechanical way. This is a fundamental tenet of the research group Nature and Life Mathematics (NaLiMa), based in Greece. This research team also support the mode of teaching based on the emergence and development of children's prior knowledge and informal methods. Furthermore, teachers should understand their students' levels and their way of thinking, in order to work accordingly. Mental calculations and estimation, as described in this book, comprise a fertile field of mathematics in which the above-mentioned views can be applied. My research team, NaLiMa, have been working in the field of mental calculation and estimation for a number of years, conducting various research and experiments in educational practice.

In their everyday lives, people are obliged to perform various calculations. Generally, these calculations are performed by three means: (a) mechanical means, namely calculator, computer, cell phone, etc., (b) paper and pencil, for executing the written algorithms of operations or (c) by carrying out calculations in one's mind, i.e. mental calculations. Evidently, combinations of these means are also used: for example, one can use a calculator for an operation and check the result mentally. Some operations can be carried out on paper, some by using machines and some using the mind. Also, calculations can be made in the mind and notes can be taken using paper and pencil.

In elementary schools, written algorithms have traditionally had a central and important place in the curricula in relation to mental calculations. Traditionally, fluency in calculations meant (and perhaps even does so today for many teachers) good knowledge and use of formal written algorithms of operations.

2 Introduction

Approximately thirty years ago, in developed countries, the largest part of mathematics classes was devoted to the teaching of written arithmetic, i.e., the formal written algorithms of addition, subtraction, multiplication and division. These skills were considered an essential preparation for adult life; through the acquisition of these skills it was thought that students gained knowledge of the concept of numbers.

Recently, changes have occurred, further emphasising the importance of mental calculation and estimation in relation to written calculations (e.g., Thompson, 1999; McIntosh, 1990; Reys, 1984; Wandt & Brown, 1957). The main reasons for this change in the approach to the two types of calculation and the emphasis on the importance of mental calculations and estimation are as follows:

First, the influence of factors of the wider social environment will be discussed. Research on adults concerning the type of calculations undertaken in their daily lives, shows that most calculations take place mentally. Wandt and Brown (1957) conducted one of the first studies in this area, at a time when the use of computers was not widespread, and found that three-quarters of the calculations made by adults were made mentally. In a more recent study by Northcote and McIntosh (1999), which examined the type of calculations used by adults during a typical 24-hour period, it was found that 84.6% of the calculations were made mentally, 11% in writing and 6.8% with calculators. Sixty per cent of all calculations were estimations and 40% were precise calculations. Regarding everyday subjects for which calculations were necessary, it was found that most calculations (24.9%) concerned time, while a second reason for calculation (22.9%) was shopping in the market. Another factor in the social environment is the widespread dissemination of cheap calculators and, more generally, the digital environment today, which increases the need for mental calculations as a means of checking answers provided by the calculator.

The second group of factors which influenced the shift in emphasis from written to mental calculations concerns the pedagogical and didactic methods and principles adopted by programmes implemented in schools. The philosophy of the programmes involves a general move toward constructivist approaches to teaching mathematics, which, as regards calculations, are expressed with the interest in understanding the strategies used by the students themselves in order to calculate mentally. Modern curricula also refer to the important concept of *numeracy* (see Chapter 1). According to this concept, people should be able to easily cope with operations in their daily life. They should be able to perform operations during the problem-solving stage and understand the importance of these operations, as well as the solutions to a given problem.

Today, it is widely known that students make many errors in lengthy written multiplications and divisions, but also in the application of algorithmic rules for operations with fractions, decimals and percentages. The reason for these errors, as well as for forgetting and confusing these algorithmic rules, is the lack of understanding. New concepts in the curricula provide opportunities for children to abandon rules and focus on understanding, so as to feel that they can control and use numbers correctly. Following the logic of the new curricula, students are asked to develop a 'number sense' with skills of knowledge and understanding and to use and apply mathematics in problem solving, communication and reasoning. It is expected that students will find meaning in problems with numbers and recognise the operations necessary to resolve them.

This book consists of seven chapters, here briefly summarised.

In Chapter 1, general theoretical subjects of mental calculations are presented. First, in Section 1.1, an attempt is made to identify mental calculation in the modern teaching environment. A definition of mental calculation is presented and the term *mental calculation* is distinguished from *estimation*, which are often confused. The reasons why mental calculation is important are explained, along with a presentation of its position in the modern curricula and its relationship to the concept of numeracy.

Section 1.2 provides an analysis of the relationship between the concept of number sense and mental calculations and estimation. Moreover, the connection between mental calculations and conceptual and procedural understanding is analysed.

In Section 1.3, an attempt is made to show the trajectory of the development of numerical abilities, from the first numerical concepts to more complex multiplicative concepts and rational numbers. This evolution of concepts and numbers aims to demonstrate the development of mental calculations. Then, we present and analyse mental strategies and their characteristics. Finally, the characteristics of mental calculations and written algorithmic operations are analysed and compared.

In the last section of Chapter 1 (1.4), the concept of flexibility and the three variables that determine it are analysed and presented, namely: (a) task variables or the characteristics of the problem, (b) subject variables or speed and efficiency and (c) context variables, relating to the socio-cultural context.

In Chapter 2, issues related to the teaching of mental computation are considered. More specifically, in Section 2.1 results of international research on the teaching of mental computation and flexibility are presented.

In Section 2.2, curricula on mental calculations are analysed and compared, specifically, the cases of England and the Netherlands, whose educational systems are considered to be among the most modern.

Chapter 3 is devoted to mental addition and subtraction, the strategies used in them, as well as research conducted.

In Section 3.1, all strategies that can be used in addition and subtraction with numbers up to 20 and the three basic levels of development of these strategies are presented. Research results for mental addition and subtraction with numbers up to 20 are presented in Section 3.2.

Section 3.3 involves all the strategies and their categorisation for mental additions and subtractions with numbers from 20 to 100. Section 3.4 focuses on studies that examine solving subtractions by using addition strategies. In Section 3.5, research from the School of Nature and Life Mathematics on mental addition and subtraction of two-digit and multi-digit numbers are included.

Chapter 4 deals with the issue of mental multiplication and division. In the first section (4.1), the issue of multiplication tables is discussed in detail. The mechanisms of learning and strategies for calculating the multiplication tables along with various factors affecting these strategies are analysed. Also, proposals and teaching instructions for teaching multiplication tables are presented.

Section 4.2 refers to the simple operations of division. Students' individual knowledge of division before teaching and the strategies that children use to carry out simple divisions are demonstrated, along with research conducted in the simple divisions. Section 4.3 presents mental calculations with multi-digit multiplication and division. Initially the categorisation of strategies used in such operations is made, and the results of research from Greek students of elementary and middle schools are introduced. The last section, 4.4, develops the subject of Greek multiplication. The historical origins of Greek multiplication and its application in education are presented.

In Chapter 5 the theme of mental calculations with rational numbers is developed. In Section 5.1, a general account of rational numbers is provided (e.g., what rational numbers are, what fractions are) along with the difficulties encountered by students in the transition from natural to rational numbers. The various strategies used by students in mental calculations of rational numbers are analysed in Section 5.2, where many examples are presented.

In Section 5.3, the results from research on rational numbers are included. There are studies on students and teachers that focus on the comparison of fractions and ultimately on instructional intervention. Students and teachers face difficulties in mental calculations of rational numbers, as they mostly use rule-based strategies rather than number-sense strategies. Finally, Section 5.4 presents research results from the School of Nature and Life Mathematics concerning mental calculation with rational numbers on the part of Greek students.

Chapter 6 refers to computational estimation. Initially, Section 6.1 shows the various types of estimation including computational estimation. In Section 6.2 the various strategies that can be used in computational estimation problems are presented. Section 6.3 presents the factors influencing the ability of computational estimation and strategy use.

In Section 6.4, the findings of research on students' and teachers' abilities in computational estimation from the international literature are presented. In Section 6.5, the results of two studies in computational estimation conducted by the School of Nature and Life Mathematics are discussed. The first research was conducted with students in fifth- and sixth-grade primary school and the second involved in-service teachers. Finally, Section 6.6 includes research findings in relation to instruction and instructional interventions in computational estimation.

Chapter 7, the last, refers to the issue of mental calculations concerning students with learning disabilities or difficulties in mathematics. In Section 7.1 the relevant terms, such as difficulties in mathematics, dyscalculia, dyslexia and the relationships among them are identified. Section 7.2 is dedicated to the memory function in order to clarify the way mental calculations are conducted. The function of working memory and the central executive system, which play an important role in the execution of mental calculations, are more extensively analysed. Section 7.3 presents research results from the fields of neuroscience concerning brain regions that determine a person's behaviour with regard to the concepts of arithmetic, especially those related to mental calculation. Section 7.4 refers to number facts and mechanisms stored in memory and the effects on students with learning difficulties caused by the dysfunction of these mechanisms.

Section 7.5 discusses how students with difficulties in mathematics use and develop strategies. Finally, Section 7.6 refers to teaching methods that are appropriate for students with learning difficulties. Research in relation to mnemonic strategies for teaching math concepts and especially multiplication tables is discussed.

References

McIntosh, A. (1990). Becoming numerate: Developing number sense. In S. Willis (Ed.), *Being numerate: What counts?* (pp. 24–43). Melbourne: Australian Council for Educational Research.

Northcote, M. & McIntosh, A. (1999). What mathematics do adults really do in everyday life? *Australian Primary Mathematics Classroom, 4*(1), 19–21.

Reys, R. E. (1984). Mental computation and estimation: Past, present and future. *Elementary School Journal, 84*(5), 546–557.

Thompson, I. (1999). Getting your head around mental calculation. In I. Thompson (Ed.), *Issues in teaching numeracy in primary schools* (pp. 145–146). Buckingham: Open University Press.

Wandt, E. & Brown, G. W. (1957). Non-occupational uses of mathematics: Mental and written – approximate and exact. *Arithmetic Teacher, 4*(4), 151–154.

1

THEORETICAL ISSUES CONCERNING MENTAL CALCULATION[1]

In this chapter, the general theoretical subjects of mental calculations and computational estimation are presented. First, the definition of terms and the definition of mental calculation and estimation are provided, along with their significance. In addition, the status of mental computation and estimation in curricula and the context of numeracy in modern teaching environments are dealt with. The relationship of mental calculation and estimation with conceptual understanding and number sense is demonstrated, including a brief demonstration of the emergence and evolution of the first numerical skills with emphasis on mental calculations. Moreover, the issue of strategies and in particular, informal strategies of children is developed. Finally, we face the issue of flexibility in mental calculations.

1.1. Mental calculations in the modern instructional environment

1.1.1. Defining mental calculation: specification of terms

In international literature, several studies have been conducted and much research has been done on mental calculation. The first interpretations and definitions of the term 'mental calculation' were determined by a lack of pencil and paper as the means of calculation. As mentioned in the introduction, calculations are considered to be made with three means: (a) the calculator, (b) mentally and (c) with pencil and paper. Wandt and Brown (1957) differentiate between 'mental' and 'paper-and-pencil' mathematics, considering mental calculation as a process of calculating a precise numerical result without the help of some external means of calculation or script. Trafton (1978) defines mental calculations as 'non-standard algorithms for computing exact answers without the use of pencil and paper'.

Later, Reys (1984, p. 548) gave a definition of the term mental calculation, making reference to estimation as well, defining the correlation between the two types of calculation.

> There are two distinguishing characteristics of mental computation. It produces an exact answer, and the procedure is performed mentally, without using external devices such as pencil and paper. Mental computation is an important component of estimation in that it provides the cornerstone necessary for the diverse numeric processes used in computational estimation.

Furthermore, in order to determine mental calculations, the concept of strategies is inserted (Thompson 1999a, 2001; Threlfall, 2002). Threlfall (2002, p. 30) features and evaluates the various ways of responding to a problem and considers strategies to be of vital importance to the wider needs of mental calculation. He considers strategies to be procedures through which a sequence of numerical modifications of a problem occurs in order to find a solution (for details see Section 1.3.2). Moreover, Thompson (2001, p. 75) argues that the phrase *mental calculation* is employed in official documents in England and Wales to highlight the importance of applying strategies in mental work.

In the same direction, Anghileri (1999) uses strategies to define mental calculation, but she also refers to 'jottings' that can be conducted during mental calculations. More specifically, she notes: 'in stressing the importance of "strategic mental methods of computation" (SCAA, 1997), distinction is made between mental recall and mental calculation (see Thompson, Chapter 12, ibid.). Mental strategies may require pencil and paper for "jottings" to support pupils' short-term memory as "mental" is interpreted as calculating "with the head" and not solely "in the head". For multiplication and division it is important that pupils remember number facts but they must also be able to derive further facts and see connections that will help them to simplify calculations' (p. 186).

From what has been mentioned above, as well as the definition of mental calculations, we can conclude that calculations are mental procedures mainly based on the use of strategies, including simple retrieval of numerical facts. They are exact calculations and constitute a part of computational estimation. Also, regarding the means of calculation, expressions such as 'without the use of any external instrument' or 'only with the mind' can be very restrictive. We can accept that sometimes in mental calculations, paper and pencil are used for 'jottings' facilitating the short-term memory. Taking all this into account, we can formulate the following definition for mental calculation:

> Mental calculation is calculation done mentally and using strategies. It produces a precise answer. Usually it takes place without the use of external media such as paper and pencil, although it can be done with a paper and pencil, to make 'jottings' that support the memory.

In the English-language literature the use of the term 'mental calculation' can be met with in such guises as: mental computation or mental calculation or mental arithmetic. We also find the general term *mental maths*, which is commonly used to express the part of mental calculations within the modern curriculum.

It is true that the two English nouns *calculation* and *computation* are usually conflated. The term calculation is assigned the broader and a more liberal significance of calculation, while the term computation is attributed a more scientific and algorithmic meaning. These two terms are usually not distinguished in the literature of mental calculations (Maclellan, 2001).

We should mention here something that confuses many teachers, namely the relationship of mental calculations with estimation. Many teachers have a general idea that mental calculations are calculations that are not the typical written ones, and are approximate. They match the concept of *mental calculation* with that of *computational estimation*.

When we execute a computation, we process numbers, the properties of the number system, number facts that we know, etc. in order to obtain an answer. The answer may be specific and unique, or it can be an approximate value. When the goal in a mental arithmetic problem is to give an exact answer, a calculation is required. The ability of this calculation to be mental depends on the size of the numbers that are involved in the operation. In contrast, if the goal is an approximate answer, or even if the numbers are great, then an estimation is required (Sowder, 1988).

According to Sowder (1988), 'estimation is the process of converting from exact to approximate numbers and mentally computing with those numbers to obtain an answer which is reasonably close to the result of an exact computation'. We can say, that is, that estimation is something more than the mental calculation that includes it. Therefore, a mental calculation is a process of conducting arithmetic operations, in order to achieve either an exact answer (in this case mental calculation is required) or an approximate answer (in this case computational estimation is required) (Maclellan, 2001).

1.1.2. The reasons why mental calculation is important

One thing noted by many researchers, that has become a common assumption, is that, mental calculation plays a significant role in the teaching and learning of mathematics.

For example, Reys (1984, p. 549) cites the following five widely accepted causes for teaching mental calculation:

1. It is a prerequisite for the successful development of all written arithmetic algorithms.
2. It promotes greater understanding of the structure of numbers and their properties.
3. It promotes creative, independent thinking and encourages students to find ingenious ways of handling numbers.

4. It contributes to the development of better problem-solving skills.
5. It is a basis for the development of computational estimation skills.

Ian Thompson (1999b, p. 147) underlines four basic reasons why mental calculation must be taught:

1. Mental calculation is more commonly used in everyday life than written calculation.
2. Practising with it leads to the creation of a better and deeper understanding of number concepts (McIntosh, 1990; Sowder, 1990).
3. Mental work develops problem-solving skills.
4. It assists in the understanding and development of written methods for calculating.

Summarising and extending the main reasons why mental calculation is important and should be taught, we shall mention the following:

A. *Usefulness and application in practice*. They are used much in everyday life and even more than written calculation.
B. *Their contribution to other mathematical concepts*. Practise with mental calculation creates a better and deeper understanding of number sense. It helps the understanding and development of the methods of written calculation. It constitutes the basis on which the capacities of computational estimation are developed. This mental work develops problem-solving abilities.
C. *Their contribution to cognitive abilities*. Mental calculation contributes to the practise in representational skills and the use of abstract concepts in short-term memory, as well as flexibility (the concept of flexibility is presented in Section 1.4). Finally, the student's metacognitive ability is exercised in the presentation of their methods of calculation.

1.1.3. Modern international programmes and mental calculations

Over the last two decades in many countries in Europe and America, major changes in the mathematics programme for primary and secondary education have been made. All this work has changed the misconception that intense training and the mere application of basic properties is the only path; in this sense classes in communities where students explore and discuss mathematical concepts are found to be in need of conversion and adjustments. An important aspect of these changes is the reduction of the practise of basic number facts and the learning and application of written algorithms. While arithmetic continues to be the focus of elementary school mathematics, the contents to which priority is given are problem solving, number sense and mental calculations. Greater attention is also paid to contents such as geometry, algebra and data handling, to which little special emphasis has traditionally been given.

In 1999, for example, the programme of a National Numeracy Strategy was introduced in primary education in England. In this programme, mental calculation plays a key role and a structured approach to teaching the strategies of mental calculations is proposed. Specifically, the direct teaching of mental calculation strategies is proposed as a whole-class approach (DfEE, 1999; DfES, 2007).

In the USA, the National Council of Teachers of Mathematics (NCTM, 1989, 2000) categorises pencil and paper algorithms as a content to be given less emphasis, while mental calculation, computational estimation and the use of calculators are to be given more attention.

Other examples of such programmes from the international arena, which include and highlight mental calculation, are: in the Netherlands, the *Dutch Specimen of a National Program for Primary Mathematics* (Treffers & de Moor, 1990); in Australia, the *National Statement on Mathematics for Australian Schools* (Australian Education Council, 1991), etc.; in Greece, mental calculation is being explored as part of the Cross Curriculum Framework (C. C. F. Δ. E. Π. Π. Σ., 2003).

1.1.4. Numeracy and mental calculations

In recent years, a concept of numeracy or numerical literacy has been developed that emphasises the relationship between the mathematics of everyday life and school mathematics; this has affected curricula in many places, such as in England, which also bears its name: the National Numeracy Strategy (DfEE, 1999; DfES, 2007). Mental calculation is connected with the concept of numerical literacy, because they both consist of contents of mathematics that are applicable in daily life. Further on, we will briefly refer to the concepts of *numeracy* and *mathematical literacy*, and we will determine the position of mental calculation in correlation with these concepts.

The concept of *numeracy* or *numerical literacy* was developed within the wider concept of *literacy*. Societies have changed and the sense of what it means for someone to be *numerate* and *literate* has also changed. Modern society in developed countries is dominated by technological development. Mathematics is the basic structure on which such technological development is based, because computer software based on algorithms has been produced, and the functional part of computers was designed based on mathematical logic. We live in the era of information and the need for information management. Calculators and computers perform much of the calculations which, in earlier times, were compulsory for human beings to carry out. Our world is dominated by the existence of changing technology, apart from teaching methods and the content of school mathematics curricula.

Nowadays, for a citizen of a developed society, the requirements in mathematical and numerical knowledge have greatly increased; it is different from past times. Paulos (1998) calls the ignorance of basic numerical knowledge *innumeracy*; he believes that this may be a serious weakness in many areas of human activity, whether in home and in private life or at work, or even in public and professional occupations.

However, the knowledge of sterile formulas and rules gained from school is not enough: it cannot imply automatic application in everyday life to the point that a person can face his/her daily needs in mathematics and be considered mathematically numerate. If someone knows, i.e., the algorithms of the four operations from school, it does not mean that this knowledge is enough to meet the demands of calculations and estimations that are required by the market. Below is presented one of the first definitions of being *numerate*, widely reported by teachers of mathematics, appearing on the Cockcroft Report of the British government, which was a milestone in redefining the goals of mathematics education in England (Cockcroft, 1982, § 34):

> We would wish the word 'numerate' to imply the simultaneous existence of two attributes. The first one is an 'at-homeness' with numbers and the ability of someone to make use of mathematical skills that enables an individual to cope with the practical mathematical demands of his everyday life. The second one is the ability of a person to appreciate and understand the information that is presented in mathematical terms, for instance, in graphs, charts or tables or by reference to percentage increase or decrease. Taken together, these imply that a numerate person should be expected to be able to appreciate and understand some of the ways in which mathematics can be used as a means of communication.

According to the above, we can conclude that arithmetic as mathematical content is something different from numeracy, and generally mathematics and mathematical literacy are different areas that can be considered complementary. Math involves abstract mental objects, products of human deductive thinking (numbers, geometrical shapes, equations, and so on). It uses mathematical proof as a means of verification and evaluation and is presented as a specialised activity. Often logic in mathematics differs from everyday logic. Mathematical literacy concerns the functional dimension of mathematical knowledge and their use in daily life activities. A definition of mathematical literacy developed by a team of experts in mathematics of the PISA programme (Program for International Student Assessment) follows (OECD, 1999):

> Mathematical literacy is an individual's capacity to identify and understand the role that mathematics plays in the world, to make well-founded judgments, and to engage in mathematics in ways that meet the needs of that individual's current and future life as a constructive, concerned and reflective citizen.

When we speak of mathematical literacy and mathematics applied in the practice of life, the situations and phenomena involved are manifold; it is often difficult to distinguish the mathematical concepts involved. What are these concepts and how can they be organised? Four phenomenological categories are suggested: *Quantity*, *Space and Shapes*, *Change and Relationships* and *Uncertainty* (Steen, 1990; OECD, 2002).

De Lange (2003, p. 79) presents the phenomenological notion of quantity as follows:

> Quantity. This overarching idea focuses on the need for quantification to organize the world. Important aspects include an understanding of relative size, recognition of numerical patterns, and the ability to use numbers to represent quantifiable attributes of real-world objects (measures). Furthermore, quantity deals with the processing and understanding of numbers that are represented to us in various ways. An important aspect of dealing with quantity is quantitative reasoning, whose essential components are developing and using number sense, representing numbers in various ways, understanding the meaning of operations, having a feel for the magnitude of numbers, writing and understanding mathematically elegant computations, doing mental arithmetic, and estimating.

This shows that mental arithmetic and estimation are explicitly part of numeracy, within the context of the initial concept of quantity. Mental arithmetic and estimation are key components of quantitative reasoning, together with the understanding of the meaning of operations, the number sense and the representation of numbers in different ways. These concepts are contained and accentuated within the modern curriculum, in contrast to older programmes, which were more traditional and more focused on mathematical concepts and their sequence. Mental calculation and estimation, even though they are very useful in everyday life, are not included in traditional courses, because they do not possess pure mathematical existence and generalisability. Most strategies for mental computation and estimation, as will be seen below, have an idiosyncratic origin and are not applicable in all cases, meaning that they do not have the ability to be generalised that is characteristic of mathematical formulas.

1.2. Number sense, mental calculation and computational estimation

1.2.1. What is number sense?

It has been more than half a century since the concept of number sense was delimited (Dantzing, 1954) to the abilities to use and understand numbers, arithmetic relations and operations. Nevertheless, the sense of number is difficult to define, so various researchers have identified several combinations, including a conceptual understanding of the number and specific numerical skills as components of number sense (Berch, 2005).

Number sense refers to a person's general understanding of numbers and operations and his/her ability to handle daily-life situations that include numbers. This ability entails the usage of useful, flexible and efficient strategies in mental calculation and estimation, to handle numerical problems (McIntosh, Reys, Reys, Bana & Farrell, 1992; Reys & Yang, 1998; Sowder, 1992). To quote McIntosh et al.

(1992), 'Number sense refers to a person's general understanding of number and operations along with the ability and inclination to use this understanding in flexible ways to make mathematical judgments and to develop useful strategies for handling numbers and operations' (p. 3).

McIntosh et al. (1992) and Dunphy (2007) refer to the substitution of the term 'numeracy' for the term 'number sense'. McIntosh et al. (p. 3) state that the word 'numeracy' was replaced by another, less abstract one not linked to a conservative and sterile view of mathematical needs. According to Dunphy (2007, p. 7) the term 'number sense' replaced the word 'numeracy' due to a desire amongst reformers to attempt to refocus on constructivist principles in mathematics education.

Koachman, Moss and Case (2001), note that 'the characteristics of good number sense include: (a) fluency in estimating and judging magnitude, (b) ability to recognise unreasonable results, (c) flexibility when mentally computing, (d) ability to move among different representations and to use the most appropriate representation' (p. 2).

Dunphy (2007, p. 11) reviewed and summarised the literature on number sense, characterising it as follows:

- a holistic construct that is difficult to define;
- concerned with the development of a wide range of understandings, skills and attitudes about number that extend beyond those generally associated with numeracy and encompass everyday uses;
- manifested in the ability to think flexibly about number;
- closely related to the development of numeric disposition;
- developed as a result of participation in everyday experiences with and about number.

According to the above, we see that mental calculations and estimations are a subset of number sense and contained therein. We can say, then, as we will show in detail below, that mental calculations and estimations are related to and depend directly on each person's number sense. Moreover, as we saw above, among the reasons why mental calculations are important (1.1.2.) is that training with mental calculations creates a better and deeper number sense.

Obviously, number sense is determined differently depending on the child's age and the knowledge possessed. So number sense in very young children may well look different from that of older learners (Carpenter, 1989; Dunphy, 2007). McIntosh et al. (1992, p. 3) state that the acquisition of number sense is a gradual, evolutionary process and begins to develop long before formal schooling.

1.2.1.1. What are the components of number sense?

In the literature we can find many studies that analyse and define a framework or the components of number sense either generally (e.g., McIntosh et al., 1992) or depending on the students (age, disabilities or not, etc.) (e.g., Berch, 2005) and the kind of numbers and the operation examined (magnitude of numbers, whole numbers, fractions, etc.) (e.g., Markovits & Sowder, 1994).

McIntosh et al. (1992, p. 4) define a general framework for number sense, with three key areas:

1. Knowledge of and facility with numbers: 1.1. Sense of orderliness of numbers. 1.2. Multiple representations for numbers. 1.3. Sense of relative and absolute magnitude of numbers. 1.4. System of benchmarks.
2. Knowledge of and facility with operations: 2.1. Understanding the effect of operations. 2.2. Understanding mathematical properties. 2.3. Understanding the relationships between operations.
3. Applying knowledge of and facility with numbers and operations to computational settings. 3.1. Understanding the relationship between problem context and the necessary computation. 3.2. Awareness that multiple strategies exist. 3.3. Inclination to utilize an efficient representation and/or method. 3.4. Inclination to review data and results for sensibility.

Berch (2005, p. 334) in a work for children with mathematical disabilities, notes 30 components of number sense.

Some authors make reference to the affective issue in their characterisations of number sense (Anghileri, 2000; Dunphy, 2007; Howden, 1989; McIntosh et al., 1992). Dunphy (2007), in her framework of number sense regarding four-year-old children, includes pleasure and interest in number. This aspect relates to disposition. De Corte, Greer and Verschaffel (1996) consider disposition to be important aspect of learning in the domain of mathematics. Concerning number sense, Anghileri (2000, p. 2) states that 'It refers not only to the development of understanding but also to the nurturing of a positive attitude and confidence that have been lacking in more dated curricula.'

1.2.2. Conceptual understanding and procedural skills in mental calculations

In this section we show how the terms *relational or conceptual understanding* and *instrumental or procedural understanding* are used in mental calculations both to interpret the interaction of conceptual understanding and procedural skills and also to separate the strategies used in mental calculations on procedural and conceptual strategies.

Skemp (1976) separated and contrasted relational understanding with instrumental understanding. According to him, relational understanding is based on the understanding of concepts and their interconnection, so that the student will know what he/she (hereafter referred to as 'he' for simplicity) is doing and why he is doing it without relying simply on the application of rules (rules without reason). In instrumental understanding, the student applies an algorithmic process mechanically.

Hiebert and Wearne (1996) developed a theory related to the influence and interaction of conceptual understanding in procedural skills. They argue that children who have adequate understanding or cognitive structures are in an advantageous position to acquire procedures and use them appropriately. Children can acquire procedures: (a) by discovering new procedures, either creating or adapting known procedures to solve new problems or (b) by adopting procedures presented by others. The authors suggest that in both cases conceptual understanding facilitates the acquisition of procedures.

Hiebert and Wearne argue that students with proper understanding have a greater advantage with any kind of information, because they possess a mental structure that can classify and give meaning to information. The procedures that are adopted can be connected with relevant knowledge of students and be understood. The relevant knowledge can be used for the control of validation and execution of processes in specific contexts, and for the support of memory for such processes.

Understanding may be important for the construction of appropriate discoveries or modification of procedures that already exist, and can be beneficial – not being necessary – for the adoption of procedures. As we know, a procedure can be adopted without being understood.

Finally, these authors report that students who present conceptual understanding are more likely to develop new appropriate procedures and adapt procedures they learn to new situations better than their peers. It is also possible that students who understand acquire procedures presented by others with more understanding and are likely to remember them better.

Hiebert and Wearne (1996) conducted a study focusing on the interaction between conceptual understanding and procedural skills. Its subjects were about 70 students in their first three years of school while they were learning the decimal number system and addition and subtraction with multi-digit numbers in two different teaching environments. One was a conventional teaching environment, where instruction occurred according to the textbook, and the other one was an alternative teaching environment that encourages students to develop their own procedures and to give meaning to procedures instructed by others. The results of this study showed that alternative teaching created higher levels of understanding for students; the students showing conceptual understanding were more likely than their peers to discover their own procedures and adapt old ones to the solution of new problems.

Heirdsfield and Cooper (2004) also conducted a study with seven third-grade students with low performance in mental calculations. They tested the students' knowledge through personal interviews and found that lack of procedural understanding is a cause of errors in mental calculations and the inability to use alternative strategies.

McIntosh et al. (1994), based on the terms of Skemp (1976) we saw above, separated strategies that students use in operations into instrumental and conceptual. This separation has been used by researchers (e.g., Callingham & Watson, 2008; Yang, Reys & Reys, 2009) in order to rank the strategies of students in *procedural* or *rule-based strategies* and *conceptual* or *number-sense strategies*.

Strategies are characterised as *procedural* or *rule-based* when students use techniques that are learned by rote and use them without showing that they understand what they are doing. For example, in the operation 1/2: 1/4 students apply a technique in which they reverse and multiply the second fraction: $1/2 : 1/4 = 1/2 \times 4/1 = 4/2 = 2$. Instead, *conceptual* or *number-sense strategies* characterised these, which show that students understand what they are doing while performing operations. For example, in the operation, 1/2: 1/4 students represent fractions as a watch and see 1/2 as a half hour and 1/4 as a quarter of an hour. They consider that half hours have 2 quarters and give the answer 2 for the operation. Another conceptual solution would be to convert fractions into decimals: $1/2 : 1/4 = 0.5 : 0.25$, 0.25 fits 2 times in 0.5.

1.2.3. Number sense, mental calculations and estimation

1.2.3.1. Number sense and mental calculations

Whitacre (2014, p. 3) considers mental computation to be a microcosm of number sense. Among the characteristics which determinate the relation of mental calculation with number sense, we can identify the two most essential: (1) The performance of mental computation reflects number sense. Literature at international levels has supported the inclusion of mental computation in mathematics curricula as a way to assist the development of number sense (e.g., Klein & Beishuizen, 1994; Markovits & Sowder, 1994; McIntosh, 1998; Reys, Reys, Nohda & Emori, 1995). (2) Mental computation contributes to number sense development. In much research concerning both students and teachers (e.g., Anghileri, 2000; Blöte, Klein & Beishuizen, 2000; Tsao, 2005; Yang, Reys & Reys, 2009; Whitacre, 2014), the results show that accurate mental computation (*accuracy*) can be the result of successfully applying efficient mental strategies (*flexibility*) that exhibit number sense.

Instead, it was found that students can succeed in calculations (*accuracy*), without understanding, meaning that high performance in mental calculations can be achieved without being accompanied by a number sense (e.g., Heirdsfield, 1996; McIntosh & Dole, 2000). These students, probably after a lot of practise, mechanically and properly use a strategy (*inflexible*). McIntosh et al. (1992) claimed that high skill in written computation is not necessarily accompanied by number sense. That is to say a student or an adult may find the correct answer to an operation mechanically, without realising the significance of 'numbers' or 'operations'. For example, as mentioned above, many students can properly calculate the division $1/2 : 1/4 = 2$ by inverting and multiplying the second fraction, but we don't know how many of them understand that a quarter can be placed twice into a half.

As we will see on the following pages in the literature review, it seems that mental calculation is influenced by and correlates with many skills and contents of the number sense, such as: understanding of the base-ten number system and its properties, the structure of numbers and their properties, the use of different representations of numbers, the knowledge of number facts and flexibility.

Reys, B. (1985) highlights that '*mental computation promotes an understanding of the base-ten number system as well as of basic number properties*' (pp. 45–46). Reys categorised strategies used by high- and middle-ability seventh and eighth-grade students. High-ability students were able to widely use different number properties, including the forms of commutativity, associativity and distributivity. The high-ability students flexibly utilised the knowledge of place value in the number system and the connections between operations. On the other side, middle-ability students tended to use the standard written algorithms and seemed unwilling to create invented strategies mentally. B. Reys (1985) states that 'mental computation nurtures the development of keen number sense' (p. 46). Additionally, R. Reys (1984) believes that mental computation 'promotes greater understanding of the structure of numbers and their properties' and 'promotes creative and independent thinking and encourages students to create ingenious ways of handling numbers' (p. 549).

Carraher, Carraher and Schliemann (1987) concluded that the oral mathematics used to solve everyday problems provides evidence of students' understanding of the decimal system. Furthermore, students tend to use strategies based upon the decomposition and composition of a number that reflect their understanding of number properties.

Reys and Barger (1994) posit that students who are good mental calculators use number properties and equivalent representations of numbers to perform transformations on numbers. Students who are less proficient tend to rely on mental versions of standard algorithms.

An ability of the number sense closely linked with mental calculations is the knowledge of number facts. The knowledge of basic number facts is a prerequisite for mental calculations and generally for arithmetical procedures (Plunkett, 1979; Sowder, 1988; Dowker, 2005). When the number facts are retrieved directly from long-term memory, working memory is facilitated and is available to calculate more complex problems (Resnick & Ford, 1981). A detailed analysis of the function of number facts in memory during mental calculations is located in Section 7.5.

Several investigations conducted (e.g. McIntosh et al., 1997; McIntosh et al., 1992; Reys, 1984; Sowder, 1990, 1992; Trafton, 1992), examined and related students' abilities in number sense with flexibility in mental calculations. The results of these investigations indicate that number sense is a basic condition for the development of flexibility in mental calculations. Many researchers argue that in mental calculations, flexibility is a component of number sense (e.g., Klein & Beishuizen, 1994; McIntosh, 1998; Tsao, 2004).

1.2.3.2. Number sense and estimation

In Chapter 6, computational estimation, one of the most common types of estimation, is presented in detail along with the other different types of estimation. As we pointed out above, the ability of estimation is an integral part of the number sense. Below we will discuss some research results that show and analyse this relationship.

Levine (1982) examined college students to identify the strategies they use in estimation. At the same time, a test was given to each student in order to examine his quantitative ability. The results showed that students with high scores on the test of quantitative ability used many and varied estimation strategies and were better in computational estimation skills in comparison with students who had lower scores in the test of quantitative ability. Levine (1982) found that those who are weak in estimation initially calculate exactly, rounding afterwards to find an estimation. She also observed that people who use this method seem to have no number sense.

Threadgill-Sowder (1984) posed 12 NAEP (National Assessment of Educational Progress) items for students in grades six through nine. This study found that 'Students who gave acceptable responses consistently demonstrated this quantitative intuition, or number sense, whereas those who gave unacceptable responses seemed to have little feel for the numbers represented' (p. 335). The results of the study indicate that estimation skills are highly dependent upon a student's number sense. Threadgill-Sowder (1984) states that good estimators, having a good understanding of basic facts, place value and arithmetic properties, are skilled at mental computation, demonstrate a tolerance for error, and can flexibly use a variety of strategies as well as displaying self-confidence.

In an article elaborating on the results of researches on estimation and number sense, Sowder (1992, p. 382), stated:

> And what about those who performed poorly on estimation and mental computation tasks? If Resnick is correct, then we must focus instruction on making sense of the symbols we use in mathematics. Instruction cannot help but develop quantitative intuition if it allows and encourages the invention of algorithms, promotes questioning of how numbers can be decomposed and recomposed and of how place-value concepts can be applied, admits to multiple answers and procedures, and demands reflection on reasonableness. Estimation and mental computation are not only useful tools in everyday life but they can also lead to better number sense.

1.3. Development of arithmetic skills, informal knowledge and students' strategies

1.3.1. The emergence and development of the first numerical skills

As we saw above, mental calculation is closely related to and is part of the number sense. That is to say, mental calculation represents a part of arithmetic skills, and it would be interesting to briefly describe the evolutionary development of numerical skills, focusing on those associated with mental calculation. Therefore, in this section, we will conduct a brief presentation of the evolutionary development of numerical concepts, starting from the first months of human life, reaching to the concepts of rational numbers.

The terms *informal mathematics* and *informal knowledge* are used not only to indicate the skills and knowledge that a child acquires in life away from school, but also the skills and knowledge it develops in school without being taught. Therefore, children, even before accepting any organised teaching in school, already have much knowledge as well as abilities and skills about numerical concepts which have a predominantly social and experimental character. The child acquires this knowledge and skills in the family environment or generally from the social and natural environment in which he lives (Gelman & Gallistel, 1978; Fuson, 1988; Resnick, 1989; Fischer, 1992; Sophian, Wood & Vong, 1995; Butterworth, 1999; Nunes & Bryant, 1996).

Informal mathematics, which younger children possess, has been extensively studied. A general conclusion, often made, is that children from different cultural and socioeconomic backgrounds progress in the same developmental course in their early and intuitive understanding of mathematics (Gelman, 2000; Ginsburg, 1982, 1997; Klein & Starkey, 1988).

Characteristically, Gelman (2000, p. 27) states that 'the cumulative strength of the evidence encourages investigators to adopt the proposal that there are invariant arithmetic structures of mind that contribute to the widespread development of an implicit but principled understanding of natural numbers. This understanding enables individuals to count and to generate cardinal values that can be used in a structure that is something like addition and subtraction.' Gelman notes that 'We come to the world of learning with a number-relevant mental structure that is comprised of skeletal principles for counting, for generating cardinal values, and for adding and subtracting the resulting cardinalities' (p. 27).

Many recent studies in cognitive, comparative and developmental psychology suggest that not only human beings but also many species of animals are born with a set of inherent and basic quantitative skills. As far as quantities are concerned, these primary abilities include an implicit understanding of cardinality, of ordinality, of counting and of addition and subtraction (increase or decrease of the quantity of small sets) (Boysen & Berntson, 1989; Geary, 1995; Butterworth, 2005).

1.3.1.1. Numerosity and counting

The concepts of *numerosity* and *counting* are important and contribute to the creation and development of the concept of number, of operations and generally the early numerical concepts for the child.

The concept of numerosity is used to show the number of objects in a set. As Butterworth states (2005, p. 3), the term numerosity used in his text as the cognitive counterpart to the term cardinality, is used in mathematics and logic.

Butterworth (2005, p. 4) notes that:

> One of the key debates is whether the child is helped to understand the special numerosity meaning by possessing a specific innate capacity for numerosities, rather than, say, a capacity for dealing with, or being sensitive to, quantities more generally. Crucial evidence comes from the people who appear to have a selective deficit in this capacity which profoundly affects their ability to learn arithmetic. This condition is known as 'dyscalculia'.

The concept of dyscalculia is examined in detail in Chapter 7.

Using the concept of cardinality and the function of counting we can find the number of elements in a set and determine the number. From the point of view of human behaviour, numerosity and counting are important; they are also linked in maths with the concepts of cardinality and set.

The four arithmetic operations (addition, subtraction, multiplication and division) can be defined in terms of operations on sets and their cardinality. For example, the addition 3 + 4 is the addition of the two cardinalities of disjoint sets with 3 and 4 elements respectively. Addition is defined by the union of two or more disjoint sets. Similarly, the other operations, subtraction, multiplication and division can be defined by set operations.

Sets and cardinality constitute a common point of communication between mathematics and reality, everyday life and by extension human behaviour. Sets are associated with reality, as they correspond to collections with many objects, while the cardinality is associated with counting as a human behaviour. So, for example, in the problem 'How many are there in total if you have 3 red plus 4 yellow roses?' – for mathematics it is the union of two sets and the sum of their cardinals 3 and 4, while for human behaviour, it corresponds to the counting of the set of 3 red flowers, and the set of 4 yellow ones and adding the two cardinals.

A child, from the first years of his life, as we shall see below, has the mechanism to estimate and count quantities. Based on cardinality and counting, he can approach numbers. He can perform functions such as: put together and make many, take away, count with units greater than 1 (e.g., 5, 10, 15), distribute into equal parts, which are informal strategies and the beginning of operations of addition, subtraction, multiplication and division, respectively. In kindergarten, the first numbers are taught, based on counting and numerosity, as well as the functions of addition, subtraction and sharing in small groups, such as put or take off one, share equally.

In first grade, based on the educational curriculum of Greece, the operations of addition and subtraction are taught with numbers up to 20, using a mental method without written algorithms. In the second and third grade, the multiplication tables and the corresponding mental division are taught. At this level, the teaching of fractions and decimals also begins. Mental addition and subtraction with large numbers are done after the third and fourth grade. Mental multiplication and division with large numbers are also done after the fourth grade. Mental calculation with rational numbers (fractions, decimals and percentages) is performed in the advanced grades of primary school and the first years of high school.

The concept of numerosity, as noted earlier, is essential for the formation of the first numerical concepts presented in the first months of human life. Many experiments have been performed on infants to examine their ability to distinguish the elements of a set. An infant from the age of four months or earlier is able to distinguish one object from two and two from three (Starkey & Cooper, 1980; Antell & Keating, 1983; Starkey, Spelke & Gelman, 1990). Starkey and Cooper (1980) first showed that infants from 4–6 months old were sensitive to the numerosity of an array of black dots.

They used a 'habituation–dishabituation' paradigm, which recognises that infants prefer novelty and will look longer at new things. Repetition causes them to habituate and lose interest, while a new thing causes them to regain interest – to dishabituate. In this study, whenever Starkey and Cooper changed the number of dots, infants would dishabituate to a new number of dots, up to four.

The possession of the concept of numerosity also means the ability to detect a change in the number of a set, when new elements are added or removed. It is, namely, the skill required for the arithmetic operations of addition and subtraction. Wynn (1992), making use of the fact that infants look longer at events that violate their expectations, showed that babies of 4–6 months are able to detect the change of addition or subtraction in numerosity.

Counting is one of the first human cultural attributes displayed and one of the most important elements of number sense in children. The counting of the number of elements of a set of discrete objects requires one-to-one matching of each counting word with an object. Therefore, to perform counting, the knowledge of the sequence of counting words of natural numbers (one, two, three, ...) is required. This ability to use counting words, which has a linguistic character, appears very early, almost with the appearance of the speech-to-child, and evolves for a long time. Studies (Fuson & Hall, 1983; Fuson, Richard & Briars, 1982) have found that the ability to use counting words starts from the age of two years, and by the end of the first grade or the beginning of the second grade children can manage to count almost up to 100.

Counting makes the first bridge from the child's innate capacity for numerosity to the more advanced mathematical achievements of the culture into which he was born. That is to say, the child counts a set of objects with more elements and finds their number. This ability of counting, which may seem simple and obvious to adults, develops in the child over the course of almost four years, starting from the age of two and progressing up to six.

Gelman and Gallistel (1978) have identified the skills, which they call 'principles', that are required to be able to count. These five principles are presented below with an example in which a child is asked to count six toy cars.

1. The child should know how to recite orally from one to six or more, holding constant the order of number words each time (the *stable order principle*).
2. To each toy car a single number word should correspond (the *one-to-one principle*).
3. The number word six, corresponding to the last toy car, shows the total number of toy cars: the cardinality (the *cardinal principle*).
4. The quantity of toy cars is not dependent on their colour, their size, whether they are the same or not, etc. (the *principle of abstractness*).
5. We can count the toy cars starting from whichever we want, taking for granted that we will count them all (the *principle of order-irrelevance*).

Butterworth (2005, p. 12) presented a table of the milestones in the early development of arithmetic. For example, a child at the age of 2.0 begins to learn a sequence of counting words (Fuson, 1992) and can do one-to-one correspondence in a sharing task (Potter & Levy, 1968). At the age of 3.6, a child can add and subtract one with objects and number words (Starkey & Gelman, 1982) and can use the cardinal principle to establish numerosity of set (Gelman & Gallistel, 1978). At the age of 6.0 a child 'conserves' number (Piaget, 1952) and at 7.0 can retrieve arithmetical facts from memory.

Butterworth (2005, p. 15) notes:

> Table 1 summarises the principal milestones in the development of arithmetic by age. There are no age norms for the milestones described here, and the ages are those at which most of the children tested demonstrate these capacities with reasonable reliability. Bear in mind that the studies described are not focused on ages, but on stages; different children can reach the milestones at very different ages.

1.3.1.2. Multiplication, division and rational numbers

As discussed above, in the curriculum, multiplication and division are introduced after addition and subtraction. The operations of multiplication and division are structured at first as addition and subtraction, thus building on the concepts of sets and numerosities, multiplication at the beginning is presented as repeated addition and division as repeated subtraction, or sharing.

From early on, children can handle situations of multiplication with the logic of matching one to many, as for example in the following: one bike has two wheels, and three bikes will have $2 + 2 + 2 = 6$ wheels. Division also occurs very early in children's development, as sharing: the concept of sharing halves from a whole or of splitting in half is met with from the first grade of primary school. Multiplicative situation studies question the common view that multiplication is nothing other than repeated addition and division is merely repeated subtraction or sharing (Nunes & Bryant, 1996). Certainly there are links between additive and the multiplicative reasoning, but there are many new concepts in multiplicative reasoning, which do not appear in additive situations. One such concept is the *ratio*, which is a new kind of sense of number, expressing a relation of one to many. The ratio does not show a cardinality and a number but a relationship between two numbers.

The problems of multiplication can be solved by repeated addition, where the quantities are *extensive* (Schwartz, 1988), namely when the quantities can be measured or counted. For example, the following are situations with extensive quantities: *How many candies are available in 4 bags, where each bag has 6 candies? How long are 8 tables together, where each table has a length of 3 metres?*

Repeated addition, however, is completely inadequate for solving problems containing *intensive* quantities. Intensive quantities represent a relationship between two extensive quantities, for example prices: *1.30 € per kilo, 15 € per hour, 9 litres per kilometre*. Due to the fact that two extensive quantities forming the intensive quantity represent different kinds of quantities – money and weight, money and time, distance and litres – children should realise that when a quantity increases or decreases, another quantity should increase or decrease, respectively. Intensive quantities, namely, express rations and therefore are situations of proportion. As Piaget (1952) pointed out, proportions are challenging.

Based on division, natural numbers and integers are extended to new types of numbers, namely fractions, decimals and percentages; the rational numbers. These numbers are very useful and are frequently encountered in everyday life, in measurement results for example. Rational numbers have properties that are not found in natural numbers and integers. For example, in the natural numbers between two consecutive numbers there is no other number (e.g., between 6 and 7), while between two rational numbers there is an infinity of numbers (e.g., between 0.6 and 0.7). In the rational numbers there is not any one number that is the exact next number, as is the case with the integers. These new features of rational numbers are points of difficulty for students. More information about the difficulties of students as regards the rational numbers is presented in Chapter 5, Section 5.1.2.

1.3.2. Strategies in mental calculation

Here we will present a general view of mental strategies and their characteristics. Mental strategies constitute a special type of cognitive processes. Although strategies are defined with small differences by various researchers, they are generally regarded as cognitive effort and goal-directed procedures being adopted to enhance the behaviour of the memory. These strategies can be tested, applied deliberately by the person and are potentially available to consciousness (Naus & Ornstein, 1983; Bjorkland & Douglas, 1997).

Strategies used to recall information from long-term memory coexist with other cognitive functions and are influenced by many factors. The most important thing of all is that strategies are developed. It has been observed that age differences affect the number of strategies available to children and the effectiveness with which they use these strategies (Bjorkland & Douglas, 1997).

Over the last thirty years, many studies have been done to record the strategies that students use in the four operations with natural and rational numbers. This work will present in detail the strategies that children use in addition and subtraction of natural numbers up to 100 (Chapter 3), multiplication and division of natural numbers (Chapter 4), in operations with rational numbers (Chapter 5) and, finally, strategies of computational estimation (Chapter 6). However, in order to provide a specific overview and a discussion over the mental strategies, we present below some samples from strategies used by children and adolescents.

First grade or preliminary school children are expected to provide the following answers for the operation of 5 + 6: One child may begin to count with the help of objects (cubes) a set of 5 cubes (1, 2, 3, 4, 5), a set of 6 cubes, and then put them together and count them all from the start. This is an initial strategy called *counting all* or *sum*. Another child may start counting from 6 and scale up to 5 more steps 6, 7, 8, 9, 10, 11. This is the strategy *counting on from larger* or *min*. A third child says that 6 = 5 + 1, which means that 5 + 6 = 5 + 5 + 1 = 11. Such a child retrieves from his memory the number facts 6 = 5 + 1, 5 + 5 = 10 and 10 + 1 = 11 and calculates from them. This child has used, namely, the constructive strategy *near-doubles*. An adolescent calculates the sum 5 + 6 with the direct retrieval of the number fact from the long-term memory.

From the third grade of preliminary school, in the operation of addition 46 + 23, we may expect various answers, including the following ones: A student starts from 46 and adds 20, summing up to 66; then he adds 3 more, reaching a total of 69. This is called the *stringing strategy* (N10). Another student says: 6 and 3 equals 9, 4 and 2 equals 6, so the total is 69. That student mentally performs the written algorithm of the addition.

Valuable information about children's strategies can be extracted from the previous examples.

A first and obvious observation is that for one and the same operation, i.e., 5 + 6, different children may use different calculation strategies. At the same time, a single child may use multiple strategies for the same operation. These strategies can be characterised according to the time they require, but they also depict the child's level of arithmetic ability, such as the use of numbers, the properties of the numbering system and the calculations, meaning the number sense. In the example of 5 + 6, the strategy *counting on from larger* is faster and more intelligent than the *counting all* strategy and the constructive strategy *near-doubles* is faster and more intelligent than either of the two before mentioned strategies. The two first strategies are based on counting and numbering respectively, while the last is based on the retrieval of arithmetic facts and calculation in short-term memory.

Students' strategies can change and conform in an operation depending on the size of the numbers, in contrast to the written algorithms, where the same method is used in all the situations. This ability to switch strategies according to the data of the operations is *flexibility*, which we will look at more closely below. For example, in the problem 46 + 23 the *stringing strategy* (N10: 46 + 20 = 66, 66 + 3 = 69) may be suitable, but in the problem 49 + 23 the *holistic strategy of compensation*, where 49 becomes 50 and 1 is subtracted from the result 49 + 23 = 50 + 23 − 1 = 72, is faster and more intelligent.

For the subject of strategies Threlfall (2002, p. 30) states:

> [A]nswers to mental calculation problems can be arrived at in different ways, not all of which are equally suited to the longer-term purposes. Children can be correct:
> 1. by recall of, or 'just knowing', a number fact;
> 2. by a simple counting procedure, in which the number sequence is recited (privately) while keeping track of the count;
> 3. by making a mental representation of a 'paper-and-pencil' method (commonly a vertically represented 'sum'), and working through the procedure mentally;
> 4. by constructing a sequence of transformations of the number problem to arrive at a solution, for example, adding 36 to 28 by first adding 20 to 36 (making 56) then thinking of the remaining 8 to be added as two 4s, adding the first 4 to make 60 then adding the remaining 4 to arrive at 64 as the answer.

Threlfall mentions that any of these answers can lead to the correct result, but the fourth kind of approach, often referred as a *strategy*, is usually considered of vital importance for the broader needs of the mental calculation. Thompson (1999a, p. 2) defines *strategies* in mental calculations in the following way:

> [M]ental strategies are more about the application of known or quickly calculated number facts in combination with specific properties of the number system to find the solution of a calculation whose answer is not known. They also incorporate the idea that, given a collection of numbers to work with, children will select the strategy that is the most appropriate for the specific numbers involved.

Research shows that the same operation, in the same circumstances, may be processed differently by different people. For this reason, many students use different strategies for the same problem given to them on two different days (Siegler & Shrager, 1984). They may also use different strategies in order to solve the same problem even if it is given to them twice the same day (Siegler, 1995; Wilkinson, 1982).

1.3.3. Written algorithms and their differences from mental strategies

In continuance, we will analyse and present the function of the written algorithms of the four operations with natural numbers and their relation with mental calculations. This is a vital and contentious issue with a great impact on teaching. Many writers that analyse the function of written algorithms (Plunkett, 1979; Thomson, 1997; Usiskin, 1998) refer to their positive characteristics, such as:

- They consist in a traditional context of basic mathematics taught all over the world for many years now.
- They are very powerful in solving problem categories, especially when calculations include many numbers and the memory may be heavily charged.
- They are automatic and can be taught and executed by someone without needing from him to analyse the base on which the algorithm is based on.
- They are quick and lead directly to an answer.
- They offer a written record of the calculation, allowing teachers and students spot errors.
- They consist an authentic, permanent and steady (unchanged) procedure, which is used for all the numbers: digit or multi-digit, whole or decimal.

Even though the above-mentioned characteristics of the algorithms seem to be strong arguments for algorithms being taught traditionally and continuing in teaching even today, there has been an intense opposite reaction on this matter. A number of researchers find that the teaching of algorithms of operations with natural numbers in the primary education is dangerous for children and damages the formation of their computing abilities (Kamii & Dominick, 1997, 1998; McIntosh, 1998; Van de Walle, 2005, 2006). These researchers state many reasons why the use of algorithms will lead to problems, such as:

- *Written algorithms are based on digits and not on whole numbers, like mental calculations; thus they function from right to left rather than left to right, where mental calculations function.*

  In the operation algorithms digits are taken separately and one at a time. For example, in the vertical written addition 46 + 35, we add from the right to the left the digits separately (6 + 5 = 11, we write 1 and 1 the carry digit, 1 + 4 + 3 = 8). Meanwhile, in a mental strategy, for example the splitting strategy (1010), we calculate: 40 + 30 = 70, 6 + 5 = 11, 70 + 11 = 81, or the stringing strategy (N10): 46 + 30 = 76, 76 + 5 = 81. We see that students using the mental strategies, manipulate the whole numbers and they analyse them according to their place value in the number system, in contrast to written algorithms, where work is done on every digit separately, independently of their place value. As Kamii and Dominick state, algorithms 'unteach' the place value in the numeric system (Kamii & Dominick, 1998).

  Also, with the exception of the division, algorithms use numbers from the right to the left in a digital-centric way, hiding the result till the very end. The result becomes obvious after the completion of the operation.

- *Written algorithms are rigid and therefore they sometimes become unnecessarily tedious.*

  In mental calculation, strategies are adjusted each time, according to the numbers in the operations: for example, a different strategy is used for the subtraction 52 − 49 and another one for the subtraction 52 − 31. On the contrary, in algorithms the same method is applied to all the problems of a single operation. So, in the subtraction 6,000 − 15, according to the algorithm,

students are obliged to use a number of carry digits, and many of them make errors.
- *Written algorithms are taught and enforced by the teaching.*

Even though teaching methods proposed today are student-centric and are based on understanding and discovery, they are difficult to be applied in the teaching of the algorithms of the operations, as in the case of the carry digits, the relocation of a place in the second digit of the multiplication, hidden part operations, etc. For example, many teachers may have observed that, in the teaching of written division, unwillingly, they used a totally directed and teacher-centric method, even though they wanted students to be the centre of the process. There is a tendency nowadays to give explanations and use manipulatives for the production of more comprehensive written operations. Nevertheless, written algorithms are difficult to be understood by students, in contrast to their own inventive mental strategies.
- *Students make more errors with traditional algorithms than with their invented strategies.*

Much past research (Ashlock, 1972; Cox, 1974; Brown & Burton, 1978) has shown that students make many errors when they execute written algorithms. The frequent, systematic and repeated appearance of these errors can be explained through retracing the steps of the algorithm. This regularity that appears in the errors shows that students focus on remembering the steps of the algorithm rather than understanding them and gaining the number sense. The many and systematic errors, namely in the algorithms, show a deficiency of understanding on the students' part. In contrast, students understand better when they use the strategies they have discovered on their own or ones acquired from one of their classmates.

Kamii and Dominick (1997, p. 58) state that

> Algorithms are harmful to children's development of numerical reasoning for two reasons: (a) They 'unteach' place value and discourage children from developing number sense, and (b) they force children to give up their own thinking. Children's natural pattern is to think about numbers from left to right. However, algorithms require them to give up this thinking and to proceed from right to left and to treat each column as ones.

Researchers have compared the performance and behaviour of students who were not taught the traditional algorithms with students that were taught them. Students that were not taught the algorithms attended innovative programmes, such as *Everyday Mathematics* or *Investigations*, and were compared with students that attended traditional programmes that included the teaching of algorithms. The results showed that the students who were taught according to the innovative programmes – the ones that did not include algorithms of the operations – understood and problem solved better than their classmates that were taught by the traditional programmes. As far as multi-digit calculations is concerned, most research indicates that students of innovative programmes perform almost on the same level with the students of the traditional programmes or even do better (Campbell, 1996; Carroll, 2000; Fuson, 2003; Mokros, Berle-Carman, Rubin & O'Neil, 1996; Riordan & Noyce, 2001).

Research (Cooper, Heirdsfield & Irons, 1996a, 1996b; Heirdsfield & Cooper, 1996) shows the effect of the teaching of written operations on children's spontaneous mental strategies. Before the teaching of the written operations, children present a variety of effective mental strategies, whereas after teaching, children tend to use one mental strategy, which seems to reflect the written algorithm that was taught by the teacher. According to this, researchers (e.g., Kamii, Lewis & Jones, 1991; Reys et al., 1995) conclude that students must be free to formulate their own mental strategies: the understanding of the algorithms is improved if children base the construction of strategies on their personal and natural method of thinking.

1.4. Flexibility and mental calculations

1.4.1. Conceptualising and defining flexibility

The word *flexibility* is associated with quick action, change, adaptivity and ultimately efficiency. In the relevant literature, this concept is mainly called flexibility (e.g., Heirdsfield, 1998; Threlfall, 2002; Verschaffel, Torbeyns, De Smedt, Luwel & Van Dooren, 2007; Threlfall, 2009), but it is also found under the name *adaptivity* (e.g., Siegler & Lemaire, 1997). Lately, when it comes to flexibility in strategies, the term *strategic flexibility* is used. Sometimes we encounter other similar terms, such as: *flexible and adaptive use of strategies* (e.g., Heinze, Star & Verschaffel, 2009) and *strategic thinking* (e.g., Craig, 2009). Lately, the term *adaptive expertise* has been used, which we will analyse below.

Some authors consider the terms of flexibility and adaptivity as synonymous and use them together (e.g., Verschaffel, Luwel, Torbeyns & Van Dooren, 2009; Heinze et al., 2009). Others, especially in the beginning, used the terms flexibility and adaptivity separately (e.g., Baroody, 2003; Feltovich, Spiro & Coulson, 1997; Selter, 2009). Those who distinguish the two terms attribute to the term *flexibility* the significance *of one's ability to move between different strategies and alternate*. In contrast, they attribute to the term *adaptivity* the significance *of one's ability to use the most appropriate strategy knows, focusing on fast and correct answer*. Thus, regarding the general double term *flexibility/adaptivity*, the use of *multiple* strategies is attributed to flexibility and the choice of *appropriate strategies* is attributed to *adaptivity*.

Verschaffel et al. (2007), based on research into flexibility performed over several years by several researchers who did not always agree on what flexibility really is, what its features were and how should it be studied and interpreted, proposed the following definition: 'By an adaptive choice of a strategy we mean the conscious or unconscious selection and use of the most appropriate solution strategy on a given mathematical item or problem, for a given individual, in a given sociocultural context' (p. 19).

In the above definition we observe that the adaptive/flexible choice of a strategy can be conscious or unconscious. Because of the fact that metacognitive processes, which are defined as conscious recognition and purposeful checking, regulate the choice of the strategy, many researchers believe that this occurs with the flexibility/adaptivity of the strategy as well. In many everyday activities there is a meretricious and controlled selection of a strategy. For example, in daily life, adults solve their problems or resolve accounts by doing mental calculations, implementing a written algorithm or using a calculator. On the other hand, there is much evidence showing that for quick and simple choices between strategies for simple additions and subtractions with numbers up to 20, people's choice of a particular strategy is not derived from a deliberate estimation of the options and a conscious recognition of the factors influencing this choice, but from rather more autonomous and implicit processes (Cary & Reder, 2002; Ellis, 1997; Siegler, 1996). Verschaffel et al. (2009) and Siegler and Jenkins (1989) believe that a strategy may not necessarily be chosen rationally or performed consciously, meaning that it can be selected and executed without the involvement of any awareness.

The terms *adaptive expertise* and *routine expertise* were introduced by Giyoo Hatano in 1982 (Hatano, 1982). Hatano (2003, p. xi) argues that one of the most important issues in the psychology of mathematics education is how students can be taught curriculum objects, so that they develop adaptive expertise. He describes adaptive expertise as 'the ability to apply meaningfully learned procedures flexibly and creatively' and contrasts it with routine expertise, i.e., 'simply being able to complete school exercises quickly and accurately without understanding'.

Keith Holyoak (1991) aptly describes the distinction: 'Whereas routine experts are able to solve familiar types of problems quickly and accurately, they have only modest capabilities in dealing with novel types of problems. Adaptive experts, on the other hand, may be able to invent new procedures derived from their expert knowledge' (p. 310). Giyoo Hatano and Kayoko Inagaki, in their 1986 paper, slightly expand the above characterisation of adaptive experts: They are able to (1) comprehend why those procedures they know work; (2) modify those procedures flexibly when needed; and (3) invent new procedures when none of the known procedures are effective.

In recent years, flexibility is no longer only determined by one's ability to possess and alternate many strategies, according to the characteristics of the problem each time. The determination of flexibility has become a much more complex manner. Verschaffel et al. (2009) identify three different types of variables that influence and determine the flexibility or adaptivity:

1. *task variables* or the characteristics of the problem (i.e., the nature of the given numbers in the problem);
2. *subject variables* or speed and accuracy (i.e., how accurately and how quickly they can perform the competing strategies on the given problem, given their personal knowledge and skills in applying these strategies);
3. *context variables*, variables related to the (sociocultural) context.

1.4.2. Task variables or characteristics of the problem

The mere possession of multiple strategies does not imply the existence of flexibility. Although the knowledge of many strategies is a prerequisite for the development of flexibility, only by itself and without the efficiency of the selection among strategies in a series of similar problems, it does not recommend flexibility. Also, as we saw above, the choice of a strategy in a specific problem can be made arbitrarily and without situational awareness. Moreover, we can claim that the continuous use of a single strategy for all the problems can be, in some cases, a more flexible method than a continuous alternation of strategies.

For example, having the subtractions: 37 − 19, 37 − 24 and 32 − 28, we cannot solve them quickly and effectively using the same strategy in all of them or randomly using different strategies, without taking into consideration the characteristics of the numbers in the operations. In the first operation, 37 − 19, because 19 is close to 20 the holistic strategy of compensation is indicated (N10C): 37 − 19 → 37 − 20 = 17 → 17 + 1 = 18. In the second operation, 37 − 24, because the units of the two numbers of the operation can be easily subtracted, the strategy of splitting can be implemented (1010): 37 − 24 = (30 − 20) + (7 − 4) = 10 + 3 = 13 or the stringing strategy (N10): 37 − 24 → 37 − 20 = 17 → 17 − 4 = 13. In the third operation, 32 − 28, because the minuend and subtrahend are very close to each other, the strategy of bridging of a ten can be implemented (A10): 32 − 28 → 28 + 2 = 30 → 30 + 2 = 32 → 2 + 2 = 4.

These concerns have led researchers to search for a better and more complex interpretation and recording of flexibility. Van der Heijden (1993, p. 80) defines procedural flexibility by relating it to specific features of the situation, as follows: 'Flexibility in strategy use involves the flexible adaptation of one's solution procedures to task characteristics.' He operationalised flexibility by analysing whether children systematically use the 1010-procedure and the N10-procedure for additions and subtractions respectively in number domains up to 100. Blöte, Van den Burg and Klein (2001, p. 628) accept the same definition for the flexibility and specifically state: 'A student is considered a flexible problem solver if he or she chooses the solution procedures in relation to the number characteristics of the problems, for example, N10C for solving 62 − 29 and $A10^2$ for 62 − 24.' With the above two references, authors initially distinguish different strategies of addition and subtraction with numbers up to 100. Then, based on the analysis of the strengths and weaknesses of these different strategies in relation to certain types of problems, they define some combinations 'problem type x type strategy' as flexible and others as inflexible.

Although this method of determining flexibility is more complex than simple identification based on the variety of strategies, the definition of flexibility is based only on features of the situation, which are still deficient and problematic. As Verschaffel et al. (2007, p. 29) show, it is possible for a particular subject and under certain conditions. The choice of the strategy which Van der Heijden (1993) and Blöte et al. (2001) call 'flexible' becomes 'inflexible', and vice versa.

Theoretical issues **31**

Aside from task variables, we will further present flexibility in relation to the other two types of factors, namely the variables of subject and context.

1.4.3. Subject variables or speed and accuracy

The second group of factors referring to variables related to the subject, has been systematically investigated and modelled by cognitive psychologist Siegler and his colleagues (Shrager & Siegler, 1998; Siegler, 1996, 1998, 2000). Siegler's research team developed a simulation model known as *SCADS* (Strategy Choice and Discovery Simulation), which explains how children make adaptive choices between strategies they know and how they discover new useful strategies.

This model was applied to the way young children's knowledge in simple arithmetic sums such as 2 + 2 or 3 + 6 is developed. According to this model, if a particular strategy (e.g., counting on from the first, or counting on from the larger, or the direct retrieval of a known number fact) is chosen to solve a particular task by a particular child, then the choice this strategy basically depends on is how fast and how efficient it is for this particular task and for that particular child, in comparison to other parallel strategies that are available in the child's repertoire. SCADS, therefore, always tends to select and implement a strategy that produces the most beneficial combination of speed and efficiency for a given individual and a given sum.

Obviously, the concept of flexibility that applies to SCADS reflects a more complex and sophisticated view of the strategy-selection process. Here, flexibility is examined based on characteristics of the individual and more specifically according to the experience, knowledge and effectiveness of the individual to perform the various strategies available in his repertoire.

The model of SCADS was based on the general model of cognitive development of the *Overlapping Wave Model* of Siegler (1996).

The overlapping waves approach has three basic assumptions: (a) at each moment children have a variety of ways of thinking about most issues; (b) these various ways of thinking compete with each other, not only during the short duration of transitional periods but rather for a long time; and (c) cognitive development involves gradual changes in the frequency of these modes of thought and the introduction of more advanced modes of thought, which gradually become more and more predominant (Siegler, 1996).

The above assumptions of the model, in terms of selection strategies, are expressed as follows: It is thought that children know and use a variety of strategies in order to solve a specific problem at a given time. These strategies compete with one another and, with the increase of the age and experience, some strategies become less frequent, others become more frequent, whereas some become more frequent and then less frequent. New strategies are discovered and some older strategies cease to be used.

Data consistent with the model of overlapping waves come from a variety of areas: arithmetic, knowledge of time, spelling, reading, motor activity, learning rules, moral reasoning and scientific experiments. They are also derived from experiments at various ages such as: motor activity in infants, arithmetic in preschoolers and scientific experiments in elementary and high school children (Adolph, 1997; Siegler & Shranger, 1984; Crowley & Siegler, 1999; Siegler, 1996; Rittle-Johnson & Siegler, 1999).

It was observed that children in all these sectors use multiple strategies at a given age and that there are variations not only in each child, but also among children. In particular, it was observed at five years old that almost all children use a variety of different strategies of addition. The same was observed with subtraction, multiplication, spelling, time and memory problems. Most children use at least three strategies. Also, it was found that for individual problems the results of competition vary, so that the same child may choose one strategy one day and a different one the next (Siegler, 1987).

Research has shown that in each area, children increasingly rely on advanced approaches as they learn more about the issue. So in single-digit addition, as they learn more over time, children come to use retrieval strategies more often than time-consuming counting strategies. Also, the choice between these strategies is adaptive and made in rather different ways. The fact that the choice is adaptive is seen from the fact that a fast retrieval strategy is used in simple problems and the more time-consuming and demanding strategies are used for the more challenging problems so that the right answer is ensured (Siegler, 1986).

The overlapping waves theory suggests that learning can be profitably analysed along five dimensions: *path*, *rate*, *breadth*, *source*, and *variability*. The *path of learning* is the sequence of knowledge states, representations, or predominant behaviours that children use while gaining competence.

The *rate of learning* concerns the amount of time or experience needed for a given acquisition. The *breadth of learning* involves the range of problems and contexts to which approaches are generalised. The *source of learning* involves the causes that set learning in motion. The *variability of learning* refers to the changing set of strategies used by individual children as well as to individual differences in other dimensions (Siegler, 2005, pp. 771–772).

The method used by researchers in previous studies for determining the choice of strategies and the speed and accuracies was the *choice method*. Presented a set of problems, the subjects assessed the strategy to be used for each one, and then the mean speed and accuracy that accompany use of each strategy was calculated.

Siegler and Lemaire (1997, pp. 71–72) note that:

> Unfortunately, the estimates of strategy characteristics generated by the choice method are biased by selection effects. These selection effects involve both the problems on which strategies are used and the people who use each strategy most often. For example, if a less accurate strategy is used mainly on easy problems and a more accurate strategy is used primarily on difficult ones, the more accurate strategy may produce lower percentages of correct answers (due to its being used on harder problems) and thus appear to be less accurate. Similarly, if good students tend to rely on Strategy A and less good

students tend to rely on Strategy B, Strategy A may appear to yield superior performance even if it is no better than the alternative (for example, because the good students are more careful in executing it).

To overcome these contradictions, Siegler, and his colleagues created the model *ASCM* (Adaptive Strategy Choice Model). This is a simulation of a calculation about strategic choices and how they change with age and experience. The objective of the model is to determine how people choose adaptively between strategies, i.e., how to select one strategy among many available in a more quick and efficient manner (Siegler & Shipley, 1995; Lemaire & Siegler, 1995; Siegler & Lemaire, 1997).

In this model, four dimensions of strategic competence are distinguished, and changes in each of them can yield overall improvements in speed and accuracy:

1. The first dimension is called *strategic repertoire (which strategies are used)* and refers to various strategies used by a person to solve a number of problems in a given problem area.
2. The second dimension is called the *relative frequency (when each strategy is used)*, which involves both the relative frequencies of each strategy and the types of problems on which the strategy is used.
3. The third dimension is called the *efficiency of strategy (how each strategy is executed)* and refers to the speed and accuracy of implementation of each strategy.
4. The fourth dimension, called the *choice of strategy (how strategies are chosen)* refers to the flexibility and adaptability of the strategy choices of the individual. This dimension indicates whether one chooses the most appropriate strategy to provide a correct answer to the problem faster (Lemaire & Siegler, 1995).

To record the flexibility in this model the researchers used the *choice/no choice method*. This method requires the examination of individuals engaged in two types of situations: (a) *choice* situations, where an individual can freely choose between strategies known to solve every problem and (b) *no choice* situations, where individuals must use a specific strategy that is indicated to them in order to solve every problem. The number of no-choice situations is equal to the number of strategies used by the individual in the choice situation. Ideally, the strategies used by individuals in situations of choice are faster and more accurate than those of no-choice situations.

The researcher now has a means to assess the flexibility/adaptivity in the strategy choice of an individual in a scientifically appropriate manner, as presented below: He collects data regarding the accuracy and speed of correct answers and the time response of different strategies in no-choice situations and compares these strategies to the choice situation. An individual is characterised by a high flexibility/adaptivity if he solves every problem of the no-choice situation with the same strategy that would solve it in the choice situation with the best results in terms of accuracy and speed.

This method has been successfully applied to assess the strategy choices of children and adults in diverse mathematical domains, including solving one-step multiplication (Siegler & Lemaire, 1997) and one-step addition and subtraction (Torbeyns, Verschaffel & Ghesquiere, 2004, 2005), currency conversion (Lemaire & Lecacheur, 2001), computational estimation (Lemaire & Lecacheur, 2002) and numerosity judgement (Luwel, Verschaffel & Lemaire, 2005).

1.4.4. Context variables

The development of sociocultural research and theories as well as their involvement in issues of cognitive development indicate that the issue of flexibility/adaptability is even more complex and broader than what has been proposed by cognitive models such as those of Siegler and his colleagues that we saw above. Research into sociocultural perspectives suggests that the alternation of strategies that is performed when solving problems depends not only upon the stated variables, the problem's characteristics and the subject's variables (speed and accuracy), but also upon variables related to the sociocultural context.

Ellis (1997) criticises the Adaptive Strategy Choice Model (ASCM), noting that the sociocultural influence plays a powerful role not only in shaping the repertoire of strategies that individuals have at their disposal to solve problems, but also in the choices they make between these available strategies. He argues that the database in which the strategy choices are carried out contains more information than simply the speed and accuracy of the strategy. He says that children over time and with experience also acquire a database on cultural norms guiding cognitive performance. Choices among cognitive strategies reflect not only implicit knowledge about strategies, but also an implicit knowledge about what a given culture defines as appropriate, adaptive and wise. The latter also guide the choice of their strategies.

Ellis also argues that cognitive psychologists' point of view on choice strategy has been developed in very limited circumstances and consequently has ended up with a limited number of parameters that could possibly affect the choice of strategies (e.g., such as speed and accuracy). Cognitive psychologists have focused only on a limited set of problem situations, most of the time posing students plain operation tasks. By doing so, they overlooked the conditions of everyday life that are involved in problems and considered that the context of the problems contexts as independent from social influences. They also focused on a subset of strategies without taking into consideration cooperation with others or the use of objects and tools from given cultures. In addition, the role of cultural factors within the same culture or between different cultures was overlooked. Finally, little or no attention was paid to emotional factors.

Until the current time, studies are few that specifically address how sociocultural factors influence the choice of strategy in the area of elementary arithmetic. Nevertheless, Ellis (1997) has carried out a bibliographic analysis of directly or indirectly related investigations in which sociocultural values shape behaviour in cognitive situations. These investigations address questions of cultural difference, which may influence the content of strategy repertoires and the state of strategies within these repertoires. They also examine the choice of strategy in social contexts, including problem solving together with others.

The influence of language on the development of strategies between two different peoples and cultures is highlighted through a typical example, in this case involving Chinese and Americans. The oral articulation of word numbers in the Chinese language is very regular: for example, numbers between 11 and 20 are composed of the word 'ten' and units (e.g., the word for 'eleven' is the word for 'ten' and 'one'). Therefore counting is much easier for Chinese than for English-speaking populations, who would have to memorise the numbers between 11 and 20. A comparison between American and Chinese children in learning to count reveals that Chinese children begin to outperform American children at around age four, when children in both cultures begin to count beyond 'ten' (Miller, Smith, Zhu & Zhang, 1995). By kindergarten, it seems there is a significant difference in the distribution of strategies used by children from America and China when they add (Geary, Bow-Thomas, Fan & Siegler, 1993). Children from China prefer more to count verbally and children from America prefer to count with fingers or recall answers from memory, often causing errors. This difference in use of strategy, with the Chinese students using oral numbering from a young age, occurs due to the characteristics of the Chinese spoken numbering system. The more complex English system leads America's children to choose slower or less reliable approaches.

Ellis also documents, with research examples, that the choice between the available strategies is shaped by the relative importance attributed to social goals, as opposed to the goals of the task. Examples of such social objectives are: the desire to present a favourable side of oneself to others, to solve problems in a group, taking into account social contexts, to maintain harmonious social relations and also the great importance that attaches to a person in a task and social objectives.

Below, we present some exemplary studies, of the few that exist, in which the sociocultural perspective on strategy choice in the area of elementary arithmetic is applied.

G. Ligouras (2012), in his doctoral thesis, investigated the influence of social factors on the flexibility of strategy choice in the mental calculations of sixth-grade elementary students in Greece. The social factor examined in this study was the family context: specifically it was found that there was a statistically significant correlation between the standard of living of the family and the flexibility of students. That means that the lower the standard of living of the family was, the lower the level of students' ability to use appropriate strategies. However, the profession of the parents was not statistically significantly associated with flexibility. It also seemed that there was a significant correlation between flexibility and the educational level of the father. That is, the higher the educational level of the father was, the greater the flexibility children showed in the use of appropriate strategies. A significant correlation existed in communication between parents and students with flexibility. It was found, namely, that when there was a good climate of communication between parents and children in the house, when children expressed their opinions freely and spoke with their parents about issues that concerned them, namely issues regarding school or issues in general, they developed better flexibility skills for strategies in mental calculations. However, it was found that there was no statistically significant correlation between flexibility and the family factors of (a) parents' interest for school or (b) development of mental calculation by parents with their children at home. At first, it may seem strange that there is no correlation to the flexibility that students present, gained through the effort of the parents in order to exercise the mental calculations in their children at home. This fact, however, can be explained if we consider that mental calculation and its instruction are very recent in Greek education and are unworkable in many classes. Therefore, mental calculation and its teaching methods are unknown to Greek parents, and their involvement at home is thought to be limited to traditional frameworks remembered from their school life.

A second example is the well-known pioneer research of Nunes and associates (Nunes, Schliemann & Carraher, 1993), which clearly indicates how the sociocultural context determines the type of numerical strategies used. In this study, strategies of third graders from poor Brazilian families representing various professions were examined. For the purposes of the investigation, similar problems were given in working conditions and in the school environment. It was found that these children, at work, like street vendors, solve problems verbally and using mental calculations with few errors, while at school, as students, they solve these problems typically using written algorithms, which lead them to more and more serious errors.

Notes

1 In this book, by the term *mental calculation* we most often refer to the exact mental calculations and computational estimation together. Although they are different terms, as we will see later on, we do this for the sake of brevity and because both types of calculation are mental.
2 Here A10 is the strategy 'bridging of a ten' by subtracting: $62 - 24 \rightarrow 62 - 2 = 60 \rightarrow 60 - 20 = 40 \rightarrow 40 - 2 = 38$.

References

Adolph, K. (1997). Learning in the development of infant locomotion. *Monographs of the Society for Research in Child Development, 62*(3, Serial No. 251).
Anghileri, J. (1999). Issues in teaching multiplication and division. In I. Thompson (Ed.), *Issues in teaching numeracy in primary school* (pp. 184–194). Buckingham: Open University Press.
Anghileri, J. (2000). *Teaching number sense*. London: Continuum.
Antell, S. E. & Keating, D. P. (1983). Perception of numerical invariance in neonates. *Child Development, 54*, 695–701.
Ashlock, Robert B. (1972). *Error patterns in computation*. Columbus, OH: Merrill.
Australian Education Council. (1991). *A national statement on mathematics for Australian schools*. Melbourne: Curriculum Corporation.

Baroody, A. J. (2003). The development of adaptive expertise and flexibility: The integration of conceptual and procedural knowledge. In A. J. Baroody & A. Dowker (Eds.), *The development of arithmetic concepts and skills: Constructing adaptive expertise* (pp. 1–34). Mahwah, NJ: Erlbaum.

Berch, D. (2005). Making sense of number sense: Implications for children with mathematical disabilities. *Journal of Learning Disabilities, 38*(4), 333–339.

Bjorkland, D. P. & Douglas, R. N. (1997). The development of memory strategies. In N. Cowan (Ed.), *The development of memory in childhood* (pp. 201–246). Hove, UK: Psychology Press.

Blöte, A. W., Klein, A. S. & Beishuizen, M. (2000). Mental computation and conceptual understanding. *Learning and Instruction, 10*, 221–247.

Blöte, A. W., Van der Burg, E. & Klein, A. S. (2001). Students' flexibility in solving two-digit addition and subtraction problems: Instruction effects. *Journal of Educational Psychology, 93*, 627–638.

Boysen, S. T. & Berntson, G. G. (1989). Numerical competence in a chimpanzee (*Pan troglodytes*). *Journal of Comparative Psychology, 103*, 23–31.

Brown, S. & Burton, R. (1978). Diagnostic models for procedural bugs in basic mathematical skills. *Cognitive Science, 2*, 155–192.

Butterworth, B. (1999). *The mathematical brain*. London: Macmillan.

Butterworth, B. (2005). The development of arithmetical abilities. *Journal of Child Psychology and Psychiatry, 46*, 3–18.

C. C. F. (Δ. Ε. Π. Π. Σ.). (2003). Cross curriculum framework, pedagogical institute, ministry of national education and religious affairs, *Government Gazette* 303B/13-3-2003.

Callingham, R. & Watson, J. (2008). *Research in mental computation: Multiple perspectives*. Post Pressed. ISBN: 978-1-921214-36-3.

Campbell, P. F. (1996). Empowering children and teachers in the elementary mathematics classrooms of urban schools. *Urban Education, 30*, 449–475.

Carpenter, T. (1989). Number sense and other nonsense. In J. Sowder & B. Schappelle (Eds.), *Establishing foundations for research on number sense and related topics: Report of a conference* (San Diego, San Diego State University Centre for Research in Mathematics and Science Education), 89–92.

Carraher, T. N., Carraher, D. W., & Schliemann, A. D. (1987). Written and oral mathematics. *Journal for Research in Mathematics Education, 18*(2), 83–97.

Carroll, W. M. (2000). Invented computational procedures of students in a standards-based curriculum. *Journal of Mathematical Behavior, 18*(2), 111–121.

Cary, M. & Reder, L. M. (2002). Metacognition in strategy selection: Giving consciousness too much credit. In M. Izaute, P. Chambres & P. J. Marescaux (Eds.), *Metacognition: Process, function, and use* (pp. 63–78). New York, NY: Kluwer.

Cockcroft, W. H. (1982). *Mathematics counts: Report of the Committee of Inquiry into the training of mathematics in schools*. London: Her Majesty's Stationery Office.

Cooper, T. J., Heirdsfield, A. M., & Irons, C. J. (1996a). Children's mental strategies for addition and subtraction word problems. In J. Mulligan & M. Mitchelmore (Eds.), *Children's number learning* (pp. 147–162). Adelaide, SA: Australian Association of Mathematics Teachers, Inc.

Cooper, T. J., Heirdsfield, A. M., & Irons, C. J. (1996b). Years 2 and 3 children's correct-response mental strategies for addition and subtraction word problems and algorithmic exercises. In L. Puig & A. Guiterrez (Eds.), *International Group for the Psychology of Mathematics Education* (Vol. 20, No. 2, pp. 241–248). Valencia, Spain: University of Valencia.

Cox, L. S. (1974). Analysis, classification, and frequency of systematic error computational patterns in the addition, subtraction, multiplication and division vertical algorithms for grades 2–6 and special education classes. *Research in Education, 9*, 130–131.

Craig, A. J. (2009). Comparing research into mental calculation strategies in mathematics education and psychology. In M. Joubert (Ed.), *Proceedings of the British Society for Research into Learning Mathematics, 29*(1) 37–42.

Crowley, K. & Siegler, R. S. (1999). Explanation and generalization in young children's strategy learning. *Child Development, 70,* 304–316.

Dantzig, T. (1954). *Number: The language of science.* New York: MacMillan.

De Corte, E., Greer, B. & Verschaffel, L. (1996). Mathematics teaching and learning. In D. Berliner & R. Calfee (Eds.) *Handbook of educational psychology* (New York, Macmillan), 491–544.

De Lange, J. (2003). Mathematics for Literacy. In B. L. Madison & L. A. Steen (Eds.), *Quantitative literacy: Why numeracy matters for schools and colleges* (pp. 75–89). Princeton, NJ: National Council on Education and Disciplines.

DfEE (Department for Education and Employment). (1999). *The National Numeracy Framework for Teaching Mathematics from Reception to Year 6.* London: DfEE.

DfES (Department for Education and Studies). (2007). *Primary Framework for literacy and mathematics: Guidance paper: calculation.* London: DfES.

Dowker, A. (2005). *Individual differences in arithmetic: Implications for psychology, neuroscience, and education.* Hove, New York: Psychology Press.

Dunphy, E. (2007). The primary mathematics curriculum: Enhancing its potential for developing young children's number sense in the early years at school. *Irish Educational Studies, 26(1),* 5–25.

Ellis, S. (1997). Strategy choice in sociocultural context. *Developmental Review, 17,* 490–524.

Feltovich, P. J., Spiro, R. J. & Coulson, R. L. (1997). Issues of expert flexibility in contexts characterized by complexity and change. In P. J. Feltovich, K. M. Ford & R. R. Hoffman (Eds.), *Expertise in context: Human and machine* (pp. 125–146). Menlo Park, Cal.: AAAI Press.

Fisher, J. P. (1992). *Apprentissages numeriques.* Nancy: Presses Universitaires de Nancy.

Fuson, K. C. (1988). *Children's counting and concepts of number.* New York: Springer-Verlag.

Fuson, K. C. (1992). Relationships between counting and cardinality from age 2 to 8. In J. Bideaud, C. Meljac & J. P. Fisher (Eds.), *Pathways to number, children's developing numerical abilities* (pp. 127–149). Hillsdale, NJ: LEA.

Fuson, K. C. (2003). *Toward computational fluency in multidigit multiplication and division.* Reston, VA: National Council of Teachers of Mathematics.

Fuson, K. & Hall, J. W. (1983). The acquisition of early number word meaning: A conceptual analysis and review. In H. P. Ginsburg (Ed.), *The development of mathematical thinking.* New York: Academic Press.

Fuson, K. C., Richards, J. & Briars, D. J. (1982). The acquisition and elaboration of the number word sequence. In C. Brainerd (Ed.), *Progress in cognitive development.* (Vol. 1). Children's logical and mathematical cognition. New York: Springer-Verlag.

Geary, D. C. (1995). Reflections of evolution and culture in children's cognition: Implications for mathematical development and instruction. *American Psychologist, 50,* 24–37.

Geary, D. C., Bow-Thomas, C. C., Fan, L., & Siegler, R. S. (1993). Even before formal instruction, Chinese children outperform American children in mental addition. *Cognitive Development, 8,* 517–529.

Gelman, R. (2000). The epigenesis of mathematical thinking. *Journal of Applied Developmental Psychology, 21,* 27–37.

Gelman, R. & Gallistel, C.-R. (1978). *The child's understanding of number.* Cambridge, MA: Harvard University Press.

Ginsburg, H. P. (1982). The development of addition in the contexts of culture, social class, and race. In T. P. Carpenter, J. M. Moser & T. A. Romberg (Eds.), *Addition and subtraction: A cognitive perspective* (pp. 191–210). Hillsdale, NJ: Lawrence Erlbaum.

Ginsburg, H. P. (1997). Mathematics learning disabilities: A view from developmental psychology. *Journal of Learning Disabilities, 30,* 20–33.
Hatano, G. (1982). Cognitive consequences of practice in culture specific procedural skills. *The Quarterly Newsletter of the Laboratory of Comparative Human Cognition, 4,* 15–18.
Hatano, G. (2003). Foreword. In A. J. Baroody & A. Dowker (Eds.), *The development of arithmetic concepts and skills* (pp. xi–xiii). Mahwah, NJ: Lawrence Erlbaum Associates.
Hatano, G. & Inagaki, K. (1986). Two courses of expertise. In H. Stevenson, H. Azuma & K. Hakuta (Eds.), *Child Development and Education in Japan* (pp. 262–272). New York: Freeman.
Heinze, A., Star, J. & Verschaffel, L. (2009). Flexible and adaptive use of strategies and representations in mathematics education. *ZDM Mathematics Education, 41,* 535–540.
Heirdsfield, A. M. (1996). *Mental computation, computational estimation, and number fact knowledge for addition and subtraction in year 4 children* (unpublished master's thesis). Queensland University of Technology, Brisbane.
Heirdsfield, A. M. (1998). Flexible/inflexible: Clare and Mandy's story. In C. Kanes, M. Goos & E. Warren (Eds.), *Teaching mathematics in new times* (pp. 241–249). Gold Coast, Australia: Mathematics Education Research Group of Australasia (MERGA).
Heirdsfield, A. M, & Cooper, T. J. (1996). The 'ups' and 'downs' of subtraction: Young children's additive and subtractive mental strategies for solutions of subtraction word problems and algorithmic exercises. In P. C. Clarkson (Ed.), *Technology in mathematics education. Annual Conference of the Mathematics Education Research Group of Australasia* (pp. 261–268). Melbourne, Vic.: Deakin University Press.
Heirdsfield, A. M. & Cooper, T. J. (2004). Inaccurate mental addition and subtraction: Causes and compensation. *Focus on Learning Problems in Mathematics, 26*(3), 43–66.
Hiebert, J. & Wearne, D. (1996). Instruction, understanding, and skill in multidigit addition and subtraction. *Cognition and Instruction, 14,* 251–283.
Holyoak, K. (1991). Symbolic connectionism: Toward third-generation theories of expertise. In K. A. Ericsson & J. Smith (Eds.), *Toward a general theory of expertise: Prospects and limits* (pp. 301–335). Cambridge, Eng.: Cambridge University Press.
Howden, H. (1989). Teaching number sense, *Arithmetic Teacher,* February, 6–11.
Kalchman, M., Moss, J. & Case, R. (2001). Psychological models for the development of mathematical understanding: Rational numbers and functions. In S. Carver & D. Klahr (Eds.), *Cognition and instruction* (pp. 1–38). Mahwah, NJ: Erlbaum.
Kamii, C. & Dominick, A. (1997). To teach or not to teach algorithms. *Journal of Mathematical Behaviour, 16,* 51–61.
Kami, C. & Dominick, A. (1998). The harmful effects of algorithms in grades 1–4. In L. J. Morrow & M. J. Kenney (Eds.), *The teaching and learning of algorithms in school mathematics* (pp. 130–140). Reston, VA: National Council of Teachers of Mathematics.
Kamii, C., Lewis, B. A. & Jones, S. (1991). Reform in primary education: A constructivist view. *Educational Horizons, 70*(1), 19–26.
Klein, A. & Starkey, P. (1988). Universals of early arithmetic cognition. *New Directions for Child Development, 41,* 5–26.
Klein, A. S. & Beishuizen, M. (1994). Assessment of flexibility in mental arithmetic. In J. E. H. van Luit (Ed.), *Research on learning and instruction of mathematics in kindergarten and primary schools* (pp. 125–152). Doetinchem, The Netherlands: Graviatt Publishing Company.
Lemaire, P. & Lecacheur, M. (2001). Older and younger adults' strategy use and execution in currency conversion tasks: Insights from French franc to Euro and Euro to French franc conversions. *Journal of Experimental Psychology: Applied, 7,* 195–206.
Lemaire, P. & Lecacheur, M. (2002). Children's strategies in computational estimation. *Journal of Experimental Child Psychology, 82,* 281–304.

Lemaire, P. & Siegler, R. S. (1995). Four aspects of strategic change: Contributions to children's learning of multiplication. *Journal of Experimental Psychology: General, 124,* 83–97.

Levine, D. R. (1982). Strategy use and estimation ability of college students. *Journal for Research in Mathematics Education, 13*(5), 350–359.

Liguras, G. (2012). *The influence of social and psychological factors of 6th grade students in mental calculation flexibility* (unpublished doctoral dissertation). Department of Primary Education, Florina, University of Western Macedonia.

Luwel, K., Verschaffel, L. & Lemaire, P. (2005). Children's strategies in numerosity judgment. *Cognitive Development, 20,* 448–471.

Maclellan, E. (2001). Mental calculation: Its place in the development of numeracy. *Westminster Studies in Education, 24*(2), 145–154.

Markovits, Z. & Sowder, J. (1994). Developing number sense: An intervention study in grade 7. *Journal for Research in Mathematics Education, 25,* 4–29.

McIntosh, A. (1990). Becoming numerate: Developing number sense. In S. Willis (Ed.), *Being numerate: What counts?* (pp. 24–43). Hawthorn, Victoria: ACER (Australian Council for Educational Research).

McIntosh, A. (1998). Teaching mental algorithms constructively. In L. J. Morow & M. J. Kenney (Eds.), *The teaching and learning of algorithms in school mathematics, 1998 yearbook* (pp. 44–48). Reston, VA: NCTM.

McIntosh, A. J., De Nardi, E. & Swan, P. (1994). *Think mathematically.* Melbourne: Longman.

McIntosh, A. & Dole, S. (2000). Mental computation, number sense and general mathematics ability: Are they linked? In J. Bana & A. Chapman (Eds.), *Mathematics education beyond 2000* (pp. 401–408). Perth: MERGA.

McIntosh, A., Reys, B. & Reys, R. (1992). A proposed framework for examining basic number sense. *For the Learning of Mathematics, 12,* 2–8.

McIntosh, A., Reys, B. Reys, R., Bana, J. & Farrell, B. (1997). *Number sense in school mathematics: Student performance in four countries.* Perth: MASTEC.

Miller, K. F., Smith, C. M., Zhu, J. & Zhang, H. (1995). Preschool origins of cross-national differences in mathematical competence: The role of number naming systems. *Psychological Science, 6,* 56–60.

Mokros, J., Berle-Carman, M., Rubin, A. & O'Neil, K. (1996). *Learning operations: Invented strategies that work.* Paper presented at the annual meeting of the American Educational Research Association, New York, NY.

National Council of Teachers of Mathematics (1989, 2000). Curriculum and evaluation standards for school mathematics. Reston, VA: The Council.

Naus, M. J. & Ornstein, P. A. (1983). Development of memory strategies: Analysis, questions, and issues. In M. T. H. Chi (Ed.), *Trends in memory development research* (pp. 1–30). New York: Karger.

Nunes, T. & Bryant, P. (1996). *Children doing mathematics.* Oxford: Blackwell.

Nunes, T., Schliemann, A. D. & Carraher, D. W. (1993). *Street mathematics and school mathematics.* Cambridge: Cambridge University Press.

Organisation for Economic Cooperation and Development (OECD) (1999). *Measuring student knowledge and skills. A new framework for assessment.* Paris: OECD.

Organisation for Economic Cooperation and Development (OECD) (2002). *Framework for mathematics assessment.* Paris: OECD.

Paulos, J. A. (1988). *Innumeracy: Mathematical illiteracy and its consequences.* New York, NY: Vintage Books.

Piaget, J. (1952). *The child's conception of number.* London: Routledge & Kegan Paul.

Plunkett, S. (1979). Decomposition and all that rot. *Mathematics in School*, *8*(3), 2–5.
Potter, M. C. & Levy, E. I. (1968). Spatial enumeration without counting. *Child Development*, *39*, 265–272.
Resnick, L. B. (1989). Developing mathematical knowledge. *American Psychologist*, *44*, 162–169.
Resnick, L. B. & Ford, W. W. (1981). *The psychology of mathematics for instruction*. Hillsdale, New Jersey: Laurence Erlbaum Associates Publishing.
Reys, B. J. (1985). Mental computation. *Arithmetic Teacher*, *32*(6), 43–46.
Reys, B. J. & Barger, R. H (1994). Mental computation: Issues from the United States perspective. In R. E. Reys & N. Nohda (Eds.), *Computational alternatives for twenty-first century: Cross-cultural perspectives from Japan and United States* (pp. 31–47). Reston, VA: National Council of Teachers of Mathematics.
Reys, R. E. (1984). Mental computation and estimation: Past, present, and future. *The Elementary School Journal*, *84*(5), 546–557.
Reys, R. E., Reys, B. J., Nohda, N. & Emori, H. (1995). Mental computation performance and strategy use of Japanese students in grades 2, 4, 6, and 8. *Journal for Research in Mathematics Education*, *26*(4), 304–326.
Reys, R. E. & Yang, D. C. (1998). Relationship between computational performance and number sense among sixth- and eighth-grade students in Taiwan. *Journal for Research in Mathematics Education*, *29*, 225–237.
Riordan, J. & P. Noyce (2001). The impact of two standards-based mathematics curricula on student achievement in Massachusetts. *Journal for Research in Mathematics Education*, *32*(4), 368–98.
Rittle-Johnson, B. R. & Siegler, R. S. (1999). Learning to spell: Variability, choice, and change in children's strategy use. *Child Development*, *70*, 332–349.
SCAA (School Curriculum and Assessment Authority) (1997). *The teaching and assessment of number at Key Stages 1–3: Discussion paper no. 10*. London: SCAA.
Schwartz, J. (1988). Intensive quantity and referent transforming arithmetic operations. In J. Hiebert & M. Behr (Eds.), *Number concepts and operations in the middle grades* (Vol. 2, pp. 41–52). Hillsdale, NJ: Erlbaum.
Selter, C. (2009). Creativity, flexibility, adaptivity, and strategy use in mathematics. *ZDM Mathematics Education*, *41*, 619–625.
Shrager, J. & Siegler, R. S. (1998). SCADS: A model of children's strategy choices and strategy discoveries. *Psychological Science*, *9*, 405–410.
Siegler, R. S. (1986). Unities in thinking across domains in children's strategy choices. In M. Perlmutter (Ed.), *Perspective for intellectual development: Minnesota Symposium on Child Development* (Vol. 19, pp. 1–48). Hillsdale, NJ: Erlbaum.
Siegler, R. S. (1987). Strategy choices in subtraction. In J. Sloboda & D. Rogers (Eds.), *Cognitive processes in mathematics*. Oxford: Clarendon.
Siegler, R. S. (1995). How does change occur: A microgenetic study of number conservation. *Cognitive Psychology*, *28*, 225–273.
Siegler, R. S. (1996). *Emerging minds: The process of change in children's thinking*. Oxford: Oxford University Press.
Siegler, R. S. (1998). *Children's thinking*. New Jersey: Prentice Hall.
Siegler, R. S. (2000). The rebirth of children's learning. *Child Development*, *71*, 26–35.
Siegler, R. S. (2005). Children's learning. *American Psychologist*, *60*, 769–778.
Siegler, R. S. & Jenkins, E. (1989). *How children discover new strategies*. Hillsdale, NJ: Lawrence Erlbaum Associates.
Siegler, R. S. & Lemaire, P. (1997). Older and younger adults' strategy choices in multiplication: Testing predictions of ASCM using the choice/no-choice method. *Journal of Experimental Psychology: General*, *126*, 71–92.

Siegler, R. S. & Shipley, C. (1995). Variation, selection, and cognitive change. In T. Simon & G. Halford (Eds.), *Developing cognitive competence: New approaches to process modeling* (pp. 31–76). Hillsdale, NJ: Erlbaum.

Siegler, R. S. & Shrager, J. (1984). Strategy choices in addition and subtraction: How do children know what to do? In C. Sophian (Ed.), *The origins of cognitive skills* (pp. 229–293). Hillsdale, NJ: Erlbaum.

Skemp, R. R. (1976). Relational understanding and instrumental understanding. *Mathematics Teaching*, 77, 20–26.

Sophian, C., Wood, A. M. & Vong, K. I. (1995). Making numbers count: The early development of numerical inferences. *Developmental Psychology*, 31, 263–273.

Sowder, J. (1988). Mental computation and number comparison: Their roles in the development of number sense and computational estimation. In J. Hiebert & M. Behr (Eds.), *Number concepts and operations in the middle grades* (Vol. 2 pp. 182–197). Hillsdale, NJ, Lawrence Erlbaum Associates.

Sowder, J. T. (1990). Mental computation and number sense. *Arithmetic Teacher*, 37(7), 18–20.

Sowder, J. (1992). Estimation and number sense. In D. A. Grouws (Ed.), *Handbook of research on mathematics teaching and learning* (pp. 371–389). New York: Macmillan.

Starkey, P. & Cooper, R. G. (1980). Perception of numbers by human infants. *Science*, 210, 1033–1035.

Starkey, P. & Gelman, R. (1982). The development of addition and subtraction abilities prior to formal schooling in arithmetic. In T. P. Carpenter, J. M. Moser & T. A. Romberg (Eds.), *Addition and subtraction: A cognitive perspective* (pp. 99–116). Hillsdale, NJ: LEA.

Starkey, P., Spelke, E. S. & Gelman, R. (1990). Numerical abstraction by human infants. *Cognition*, 36, 97–128.

Steen, L. A. (1990). *On the shoulders of giants: New approaches to numeracy*. Washington, DC: National Academy Press.

Thompson, I. (1997). Mental and written algorithms: Can the gap be bridged? In I. Thomson (Ed.), *Teaching and learning early number* (pp. 97–109). Buckingham: Open University Press.

Thompson, I. (1999a). Mental calculation strategies for addition and subtraction: Part 1. *Mathematics in School*. November, 2–4.

Thompson, I. (1999b). Written methods of calculation. In I. Thompson (Ed.), *Issues in teaching numeracy in primary schools* (pp. 169–183). Buckingham: Open University Press.

Thompson, I. (2001). Issues for classroom practices in England. In J. Anghileri (Ed.), *Principles and practices in arithmetic teaching* (pp. 68–78). Buckingham: Open University Press.

Threlfall, J. (2002). Flexible mental calculation. *Educational Studies in Mathematics*, 50, 29–47.

Threlfall, J. (2009). Strategies and flexibility in mental calculations. *ZDM Mathematics Education*, 41, 541–555.

Threadgill-Sowder, J. (1984). Computational estimation procedures of school children. *Journal of Educational Research*, 77(6), 332–336.

Torbeyns, J., Verschaffel, L. & Ghesquiere, P. (2004). Strategy development in children with mathematical disabilities: Insights from the choice/no-choice method and the chronological-age/ability-level-match design. *Journal of Learning Disabilities*, 37, 119–131.

Torbeyns, J., Verschaffel, L. & Ghesquiere, P. (2005). Simple addition strategies in a first-grade class with multiple strategy instruction. *Cognition and Instruction*, 23, 1–21.

Trafton, P. R. (1978). Estimation and mental arithmetic: Important components of computation. In M. Suydam & R. E. Reys (Eds.), *Developing computational skills* (pp. 196–213). Reston, VA: National Council of Teachers of Mathematics.

Trafton, P. R. (1992). Using number sense to develop mental computation and computational estimation. In C. J. Irons (Ed.), *Challenging children to think when they compute* (pp. 78–92). Brisbane: Centre for Mathematics and Science Education.

Treffers, A. & de Moor, E. (1990). *Proeve van een nationaalprogramma voor het rekenwiskunde-onderwijs op de basisschool. Deel 2: Basisvaardigheden en cijferen.* [Specimen of a national program for primary mathematics teaching. Part 2: Basic mental strategies and written computation.] Tilburg, The Netherlands: Zwijsen.

Tsao, Y.-L. (2004). Exploring the connections among number sense, mental computation performance, and the written computation performance of elementary preservice school teachers. *Journal of College Teaching & Learning, 1*(12), 71–90.

Tsao, Y. L. (2005). The number sense of preservice elementary school teachers. *College Student Journal, 39*(4), 647–679.

Usiskin, Z. (1998). Paper-and-pencil algorithms in a calculator-and-computer age. In L. J. Morrow & M. J. Kenney (Eds.), *The teaching and learning of algorithms in school mathematics, 1998 yearbook* (pp. 7–20). Reston, VA: NCTM.

Wandt, E. & Brown, G. W. (1957). Non-occupational uses of mathematics: Mental and written – approximate and exact. *Arithmetic Teacher, 4*(4), 151–154.

Whitacre, I. (2014). Strategy ranges: Describing change in prospective elementary teachers' approaches to mental computation of sums and differences. *Journal of Mathematics Teacher Education, 18*(4), 353–373.

Wilkinson, A. C. (1982). Partial knowledge and self-correction: Developmental studies of a quantitative concept. *Developmental Psychology, 18*, 876–893.

Wynn, K. (1992). Addition and subtraction by human infants. *Nature, 358*, 749–751.

Van de Walle, J. A. (2006). *Elementary and middle school mathematics: Teaching developmentally.* 6th edition. Boston: Allyn & Bacon.

Van de Walle, J. A. (2005). Do we really want to keep the traditional algorithms for whole numbers? Draft version – copyright John Van de Walle, April 2005.

Van der Heijden, M. K. (1993). *Consistentie van aanpakgedrag.* [Consistency in solution behaviour.] Lisse, The Netherlands: Swets & Zeitlinger.

Verschaffel, L., Torbeyns, J., De Smedt, B., Luwel, K. & Van Dooren, W. (2007). Strategy flexibility in children with low achievement in mathematics. *Educational and Child Psychology, 24*(2), 16–27.

Verschaffel, L., Luwel, K., Torbeyns, J. & Van Dooren, W. (2009). Conceptualizing, investigating, and enhancing adaptive expertise in elementary mathematics education. *European Journal of Psychology of Education, 24*(3), 335–359.

Yang, D. C., Reys, R. & Reys, B. (2009). Number sense strategies used by pre-service teachers in Taiwan. *International Journal of Science and Mathematics Education, 7*(2), 383–403.

Yea-Ling Tsao (2004). Exploring the connections among number sense, mental computation performance, and the written computation performance of elementary preservice school teachers. *Journal of College Teaching & Learning, 1*(12).

2

TEACHING MENTAL CALCULATION

In this chapter, we will discuss issues in the teaching of mental calculation. In the first section, various tendencies prevailing in the international field of the instruction of mental calculation are presented along with a reference to the teaching of flexibility and adaptive expertise. In the second unit, certain typical cases of teaching in countries such as England and the Netherlands are dealt with and compared.

2.1. International research results for the teaching of mental calculation and flexibility

As already mentioned, the teaching of mental calculation is considered a major subjects in modern curricula. Therefore, we should explore how the instruction of mental calculation takes place. In the literature, there are two approaches to teaching of mental calculation. The first approach claims that strategies of mental calculation must not be taught directly but instead through a focus on students' informal strategies, either through discovery by them or through use to solve computational problems (e.g., Buzeika, 1999; Heirdsfield, 2006). The second approach considers that some specific strategies should be taught (e.g., Buzeika, 1999). Moreover, there is an attempt to give answers to questions such as: How do students learn mental calculation? Should we teach students calculation strategies directly, or, rather, not teach them and rely on strategies they find on their own? Generally, how should we teach mental calculation?

The literature on strategy comprehension shows that students usually do not employ their knowledge of strategies in the most effective way (Garcia & Pintrich, 1994). Recent research has also shown that students, despite knowing different types of strategies for mental addition and subtraction, do not frequently use this knowledge in order to solve problems (Heirdsfield & Cooper, 2004a, 2004b).

For instance, many students calculate step by step, using their fingers or mental images from standard written algorithms, whereas it should be possible to retrieve an answer from their memory (Siegler & Campbell, 1989).

Blöte, Klein and Beishuizen, (2000, pp. 228–229) attempted to answer the question 'Why does a discrepancy between strategic knowledge and implementation of that knowledge occur in a specific task?' and recorded studies seeking the factors that influence the use of strategies by students, as is illustrated below:

> Some authors link the 'production deficiency' (Flavell, 1970) to the development of children's metacognitive abilities (Borkowski & Kurtz, 1987; Campione & Brown, 1977; Flavell & Wellman, 1977). Not only do children need to know when and how a certain strategy can be implemented (Borkowski, 1985), but they also need to recognize the value of the strategy (Kurtz & Weinert, 1989) and the relative usefulness compared to other strategies (Borkowski & Kurtz, 1987; Fabricius & Hagen, 1984). Students also have to manage how much effort they put into a strategy (Pintrich & De Groot, 1990). Effort management and motivational variables have been found to play a role, next to cognitive variables, in strategy use (Ames & Archer, 1988; Garcia & Pintrich, 1994; Meece, Blumenfeld & Hoyle, 1988; Pintrich & De Groot, 1990). Moreover, an alternative view on the study of cognition and learning places a strong emphasis on the kind of setting in which a child performs. Classroom context is considered to be a very important factor in influencing students' strategic behaviour. (Garner, 1990; Resnick et al., 1992)

2.1.1. Direct teaching of strategies in mental calculations

The following question is raised: are students capable of discovering mental strategies? Do these need to be taught or should we let them develop spontaneously? England's is an example of an educational system in which the spontaneous development of mental strategies by students is encouraged (Cockcroft, 1982; DES, 1989; DfE, 1995; DfEE, 1999), even though these attempts are not always successful (Foxman & Beishuizen, 1999; Earl et al., 2000). Subsequently it was ascertained, as was mentioned before, that some students – and also some adults – do not use mental-calculation strategies effectively (Askew et al., 1997). Furthermore, it has been noted in research that high-performing students use a variety of mental calculation strategies, while students with below-average performance tend to use step-by-step calculation or typical written algorithms ineffectively (Gray, 1997; Askew et al., 1997). Gray (1997) observed that high-performing pupils are capable of establishing connections with the mathematics they already recognise and apply their knowledge to resolve novel problems. Thus, these students use mental strategies productively, which means that they can combine known number facts in order to produce new ones (Askew et al., 1997; Gray, 1991). According to these results, it appears that if productive strategies of mental calculations are not taught, some students will never have access to them. So, in England, in a discussion document issued by the School Curriculum and Assessment Authority (SCAA, 1997, p. 29), it is recommended that the development of mental calculation strategies should not be left to chance. Further, in the national English numeracy programme (National Numeracy Strategy) (DfEE, 1999, Section 1, p. 6) it is noted that some mental strategies can be developed intuitively, whereas others must be explicitly taught.

Blöte et al. (2000), in their research, have investigated the relation between mental strategies that second graders prefer to use, and the value attributed to the selected strategies by these students. The instruction these students received was within the framework of the Dutch Realistic Mathematics: they were first taught the N10 strategy (a sequential counting method), and then the 1010 strategy (a decompositional method – the particulars of these methods will be discussed further below). At the beginning of the study, children used and preferred the N10 strategy. However, after the introduction of the 1010 strategy, the use and appreciation of the 1010 strategy increased. This fact exemplifies how much the direct teaching of calculation strategies affects the way students use their strategies.

In another study, Cooper, Heirdsfield and Irons (1996) summarise research focusing on the impact of teaching mental strategies. These authors, based on the literature, indicate the need to reduce time spent on teaching standard written algorithms and to start a thorough investigation of mental algorithms. They also note that students with low performance in mental calculations must be identified and immediately taught mental strategies.

Klein and Beishuizen (1994, p. 127) also summarise studies investigating mental calculation strategies and whether these strategies should be taught. Based on the literature, the writers conclude that 'the didactic sequence should be the following: first, the N10 strategy in order to enforce mental (not in columns) arithmetic, and the 1010 strategy much later, as a transition to a written (in columns) arithmetic'. In another survey, Klein et al. (1998) summarise the literature and note that students who are less able to solve addition and subtraction problems, require structured instruction, in which the teacher helps them to construct mental strategies to solve problems.

In their research, Klein and Beishuizen (1994) observe experimental classes that differ in teaching order and educational planning. In some classes, the N10 strategy was introduced from the beginning, while flexibility was encouraged, since in this way students became more capable of using this strategy. The term 'flexibility' means, for these authors, making someone competent in the use of strategies similar to the N10 (e.g., N10C and A10). In other classes, N10 strategy and flexibility are simultaneously emphasised from the beginning, while at the same time the teaching process also focuses on informal strategies of students. The results showed that only 16 out of 176 students preferred to use the 1010 strategy when solving subtraction problems with multi-digit numbers. On the other hand, 104 students used the N10 strategy for the same problems. For addition problems, 87 students used the N10 strategy while only 49 students used the 1010 strategy. Based on survey results, it can be concluded that, if students are taught by means of a specific mental method to solve addition and subtraction problems using multi-digit numbers, they do not make an effort to invent different mental strategies.

2.1.2. Encouraging students to develop individual strategies for mental calculation

There are many surveys showing that students are able develop effective and skillful strategies spontaneously without being taught (Kamii, Lewis & Jones 1991; Kamii, Lewis & Livingston, 1993; Carpenter, Franke, Jacobs, Fennema & Empson, 1998; Heirdsfield, 2000). In their research Kamii et al. (1993) highlighted that students are able to invent their own strategies if they are not taught a particular way to solve addition and subtraction problems with multi-digit numbers. In their proposed instruction, the basic requirement is that teachers create an environment in the classroom such that students feel comfortable enough to share ideas and solutions with their classmates. Then the teachers write problems on the board and ask students to find a quick and easy way to solve them. If a solution is not found, this means that the question on the board is too difficult for the level of the class and that easier problems should be posed.

Carpenter et al. (1998) conducted a longitudinal study to investigate the development of students' understanding of multi-digit additions and subtractions. In the instruction that students received within the framework of this study, the following three features prevailed: (1) communication in the classroom, allowing students to discuss with the teacher the proposed solutions, (2) material was used in class which allowed the students to use, as long as they wanted, tens-blocks and other base-ten materials, to solve the problems and (3) the kind of problems that were introduced to the students. Word problems were given that were susceptible to groupings of multi-digit numbers. When word problems were given, students used more mental strategies than in situations wherein simple numerical exercises were given. Many surveys have come to the same conclusion on this kind of problem (Heirdsfield & Cooper, 2004a; Murphy, 2004).

These three features proved to be the most important factors contributing to the discovery of mental strategies in calculations by the students. The authors concluded that students are able to invent strategies for addition and subtraction without being taught. Furthermore, it should be noted that, if students are taught calculation strategies directly, 'there would be a danger that children would learn them as rote procedures in much the same way that they learn standard algorithms today' (Carpenter et al., 1998, p. 19).

In another investigation, Heirdsfield (2000) conducted an in-depth clinical examination of thirteen third-grade students, in order to investigate the causes leading some students to be better at mental calculation, particularly addition and subtraction, than others. In this study it was found that students who invent their own strategies are more accurate and retain more of the number sense in relation to the strategies taught by the teacher. Therefore, Heirdsfield concludes that students should be encouraged to formulate their own strategies when solving addition and subtraction problems with multi-digit numbers.

2.1.3. Theories on factors affecting learning of mental calculation

From the perspective of instruction as a participatory process, it is considered that learning occurs through participation in a community and that the classroom can be seen as a community of learners. Lave's (1988) study of the informal arithmetic of American adults showed mental strategies as active, flexible answers to problems, where the strategies used were associated with the situation. The 'place of knowledge is within a community of practice' (Lave & Wenger, 1991, p. 100) and situated in a context.

According to this theory, important factors in the study of mental calculations are considered to include the following: First, the communication and dialogue that takes place in the classroom: students show the strategies they use and discuss them with their teacher and classmates. Second, the framework within which students are asked to solve the presented problems: as was mentioned above, students tend to use mental strategies more often when they solve word problems rather than plain arithmetic tasks. The context in which we choose to present problems is also important, because it determines both the motivation of students to deal with the problem, linking school with everyday life, and also determines the type of strategy chosen by the students.

On the other hand, the constructivist perspective supports a more idiosyncratic interpretation of learning strategies, recognising the development of arithmetic based on personal knowledge.

Many studies have examined children's learning of mental calculation strategies from a constructivist perspective (Steffe, 1983; Kamii, 1985; Hiebert & Wearne, 1996; Smith, 1999).

Close to the constructivist perspective is the theoretical framework constructed by Hiebert and Wearne (1996) concerning the creation of new strategies and adopting strategies used by others. Understanding is defined as resulting from building connections or relations between representations of mathematical ideas, ideas related to quantities, the ways quantities are decomposed and the way they are regrouped and recombined.

If a processed structure of connections already exists in students' mental representations, it enables them to respond with greater flexibility to problems. This structure allows students to connect a piece of information with many other pieces or combinations of pieces, which offer them the opportunity to respond in various ways.

Many researchers attempt to combine and integrate these two theories of participation in community and constructivist perspective. Murphy (2004) notes, 'Children's use (of the strategies) may be influenced and shaped by the classroom discourse, but it may also be shaped by their previous experiences and knowledge'

(p. 15). Goos et al. (1999) emphasised the 'concept of the classroom as a community of practice' (p. 57) but also acknowledged that mathematics is created 'from reflective inner dialogue' (p. 59).

We believe that we can have a selective logic in relation to the theories, meaning that we should not follow them strictly in every situation, but combine and adapt them in accordance with the circumstances. Sfard (1998) agrees with this logic, commenting that exclusive reliance on a theoretical viewpoint can lead to 'didactic single-mindedness' (p. 11) and suggests that a model of learning should reflect a dialectic between the situation and the learner, but also take into account his previous experience.

2.1.4. Teaching flexibility and adaptive expertise

In this section we will present the topic of teaching flexibility and adaptive expertise in accordance with contemporary literature.

In many programmes in the international field, the flexibility/adaptivity of students in mental calculation is mentioned and considered as an important goal. Some such programmes are: the *Proeve van een Nationaal Programma voor het Reken/ wiskundeonderwijis* in the Netherlands (Treffers, de Moor & Feijs, 1990), the *New United Kingdom Primary Framework for Literacy and Mathematics* in England (DfES, 2007), the *Curriculum and Evaluation Standards for School Mathematics* of the National Council of Teachers of Mathematics in USA (1989, 2000), the *Handbuch produktiver Rechenübungen* in Germany (Wittmann & Müller, 1990–1992) and the *Australian National Statement on Mathematics for Australian Schools* in Australia (Australian Education Council, 1991). In all these programmes the basic idea has been that it is appropriate to encourage the students to conquer the flexibility/adaptivity in mental calculation strategies.

In contemporary literature (Hatan & Oura, 2003; Heinze, Marshick & Lipowsky, 2009; Liguras, 2012; Verschaffel, Luwel, Torbeyns & Van Dooren, 2009, 2011) the discussion on the issue of teaching flexibility/adaptivity and adaptive expertise revolves around three basic questions: When is the appropriate time to begin teaching? Who should be addressed? And how should it be performed?

2.1.4.1. When should teaching start?

Verschaffel, et al. (2009, 2011) pose and investigate the question: when is it an appropriate time to begin the effort to acquire adaptive expertise in mathematics education? Several authors (e.g., Geary, 2003; Milo & Ruijssenaars, 2002) argue that it is better if teaching aims first and above all at routine expertise and only then should the aims and pedagogy be changed toward the adaptive expertise. This claim is based on the generally accepted idea that without already-learned number facts in the long-term memory, the processes, the models and the representational tools, there cannot be flexible/adaptive thinking. It is also based on the finding that many people who have many years of experience in solving problems in a given field are unable to go beyond the routine expertise (Frensch & Sternberg, 1989; Hatano & Oura, 2003).

This perspective of a successive sequence, first routine expertise followed by adaptive expertise, as it is proposed by the authors, has been adopted by experts in mathematics education as implemented in elementary arithmetic; it is stated as follows: First, one should teach and make practical application of the use of a strategy, in order to achieve procedural fluency, and then start working on the variety and flexibility of strategies. Thus, in the initial stage of teaching and learning, a strategy is introduced to all children to solve all problems of a particular type. For example, in the first grade, the bridging through ten strategy can be taught to solve all addition and subtraction problems (e.g., $7 + 4 \rightarrow 7 + 3 = 10 \rightarrow 10 + 1 = 11$. $13 - 5 \rightarrow 13 - 3 = 10 \rightarrow 10 - 2 = 8$) or in the second grade for the solution of every subtraction with numbers from 20 to 100, in order to teach the stringing strategy (e.g., $43 - 27 \rightarrow 43 - 20 = 23 \rightarrow 23 - 7 = 16$).

Several researchers have expressed their opposition to the above point of view (Baroody, 2003; Gravemeijer, 2004; Selter, 1998; Warner, Davis, Alcock & Coppolo, 2002). They consider that the development of adaptive expertise is not something that happens after the development of the routine expertise, but the education for adaptive expertise can start from the beginning of the teaching/learning process. Adaptive expertise includes mental habits, attitudes, and ways of thinking and the organisation of knowledge that are different from routine expertise; it takes time to develop.

Verschaffel, et al. (2011, p. 189), regarding this subject, claim that:

> Applied to elementary school mathematics, for example, the latter position would imply that one does not postpone pupils' confrontation with and exploration of a variety of strategies until they have developed full routine mastery in one particular strategy, but stimulate strategy variety and flexibility already from the very beginning of the teaching and learning process. This approach is adopted in many reform-based textbooks that are becoming used more and more in Western countries, such as Germany and the Netherlands (cf. Gravemeijer, 2004; Wittmann & Müller, 2004).

We have little empirical data from studies that support teaching with strategies and have variety and flexibility. Further down we present the research of Blöte et al. (2001) who examined 206 students in the second grade on the effect of two teaching programmes on flexibility in strategies of mental addition and subtraction with numbers from 20 to 100. The first teaching programme (Realistic Program Design, RPD) teaches conceptual understanding along with procedural skills and emphasises variety and flexibility in strategies. The second teaching programme (Gradual Program Design, GPD), which is more traditional, emphasises the acquisition of basic processes, aiming at obtaining and good use of a simple basic procedure. Only at the end of the programme is attention paid to variety and flexibility in strategies.

The results of this research showed that the first programme (RPD) leads students to a higher level of flexibility than the second programme (GPD), suggesting that the first programme is better for the instruction of conceptual knowledge. However, it is found that in both student groups, real flexibility strategies were substantially delayed and conceptual understanding seemed to precede the procedural skills. It was also found that, regardless of the two programmes, children who used only one strategy for an extended period of time later had great difficulty in adopting new strategies.

2.1.4.2. Who will the teaching be addressing?

Relevant and connected to the question 'When should teaching start, to make students flexible and adaptive?' is the question: *Can all students become flexible?*

In some surveys (Hatano & Oura, 2003; Liguras, 2012; Threlfall, 2009; Verschaffel et al., 2009) this issue is raised, namely whether medium-achieving or weak students will ever be able to use strategies flexibly to reach a level of adaptive expertise. Threlfall (2002) refers to the argument often heard 'that only the more "mathematically minded" children will be capable of learning how to make good choices, so flexibility should be abandoned as an objective for the "average" and "below average"' (p. 40).

There are studies that conclude that students with low performance do not benefit from instruction based on modern programmes aimed at the development of adaptive expertise (Baxter, Woodward & Olson, 2001; Boaler, 1998; Fuson, Carroll & Drueck, 2000; Geary, 2003; Milo & Ruijssenaars, 2002; Woodward, Monroe & Baxter, 2001). On the contrary, other studies suggest that such instruction, based on modern programmes, enriches the mathematical behaviours of children with poor performance more than traditional or direct instruction, performed in relation to a specific strategy (Bottge et al., 2002; Cichon & Ellis, 2003; Klein, Beishuizen & Treffers, 1998; Moser Opitz, 2001; Van den Heuvel-Panhuizen, 2001).

Hatano and Oura (2003) also dealt with this question by treating it as an expertise subject. They describe the features possessed by the expert: he is characterised by flexibility, innovation and creative skills. He implements his own shapes in more adaptive and coordinated ways. He can understand why his procedures work, modifies known procedures and even invents new ones. He responds flexibly, when context varies. The authors note that these are the characteristics of flexibility in the calculation strategies and pose the following question: *Can all children become experts in mental arithmetic?* They report that (p. 28):

> While basic schools cannot make students real experts, they can place students on a trajectory toward expertise or prepare them for future learning (Bransford & Schwartz, 1999). In this sense, an important goal of basic schooling is to make each student a 'baby adaptive expert' of the domain or topic of choice.

Verschaffel et al. (2009, p. 346) report that more research is needed to determine whether it is indeed feasible and useful to design and implement teaching approaches that aim at flexibility/adaptivity of strategies and the success of all children, including average-achieving and especially weaker ones in mathematics.

2.1.4.3. How to perform instruction?

The third and last question of the teaching is: *How can we design and implement teaching, aiming at flexibility and adaptive expertise?*

Verschaffel et al. (2009, pp. 347–48) use an example of two textbooks – one Flemish and one German – in order to show and judge two different ways of teaching flexibility/adaptivity. The authors conclude that there is not, finally, an easy and direct shortcut to becoming flexible/adaptive and that adaptive expertise is not something that anyone can be trained in or taught. Rather it is something that has to be *promoted* or *cultivated*. Therefore, the acquisition of adaptive expertise takes place in the sociocultural context of the classroom and is an affective and cognitive matter.

Heinze et al. (2009) also conducted a study on the teaching of flexibility. These authors, in a literature study that was carried out on the empirical studies on the use of flexible strategies in multi-digit additions and subtractions, conclude that students in primary school seem to have difficulties in choosing strategies in a flexible way. An interesting interpretation of this conclusion, among other things, is that most of these studies were conducted in classrooms where traditional approaches were used for teaching arithmetic (Selter, 2001).

This survey was conducted with 245 German third-grade students in 12 different classes: every four classes, three different ways of teaching were employed, of which one was the traditional one, while the other two were consistent with the logic of modern programmes. These ways of teaching are presented below:

- The *routine approach*, in which students are taught and receive practise by means of repetition in a strategy that can be implemented quickly and correctly. Subsequently, other strategies were introduced and discussed in the classroom and their flexible and adaptive use was discussed.
- The *investigative approach*, which initially asked students to discover their own strategies. These strategies, with the help of the teacher, were reduced to a set of basic strategies and then students practised using these strategies, thereby gaining experience and a capacity for flexibility.
- The *problem-solving approach*, in which no set of key strategies is proposed, but the opportunity to create specific strategies based on existing knowledge and experience is constantly given to the students. This approach also takes into account the characteristics of numbers.

The results of this study showed that, concerning the accuracy and efficiency of students, the investigative approach and the problem-solving approach had similar results. These two groups were superior to the routine group. It seemed that the problem-solving approach, where emphasis was given to devising strategies and the rejection of practise of strategies selected by the teacher, was indeed successful in teaching an adaptive use of strategies. There was however, a large percentage (31%) of failure, wherein many exercises either were not answered or were solved by using strategies that led to incorrect results. In the investigative approach, advantages and disadvantages emerged. First, many students used strategies that led to the correct answer, even ones with low achievement. Second, two basic strategies (splitting (1010) and stringing (N10)) were chosen more often by the students taught through the problem-solving approach.

According to the above results, the problem-solving approach can be criticised: perhaps such an open constructivist approach is not appropriate for all students. On the other hand, a criticism of the investigative approach is that the training of students in selected strategies prevents them from making adaptive choices of strategies, as they tend to ignore the characteristics of the problem.

In the examination of Mathematics Education literature we find several suggestions on how to work in order to achieve flexible/adaptive strategies in mental calculation (e.g., Anghileri, 1999; Beishuizen, 2001; Buys, 2001; Thompson, 1999; Threlfall, 2002). To these proposals, regarding the creation of flexible/adaptive mental calculations, we can add those proposed in the literature on the development of adaptive expertise and its sociocultural side, in particular (Hatano & Oura, 2003; Liguras, 2012; Verschaffel et al., 2009).

Verschaffel et al. (2009) emphasise the great importance of the sociocultural dimension for the development of flexibility and adaptive expertise. They hold that research should move in this direction and consider the influence of sociocultural characteristics in the development of flexibility and adaptive expertise. Important work in this direction was done by Cobb and his colleagues (McClain, Cobb & Bowers, 1998; Yackel & Cobb, 1996). These researchers conducted a series of studies in the classroom, where an emphasis was put on the appropriate culture of the classroom for arithmetic lessons from a socioconstructivist perspective from the beginning of elementary mathematics education.

Verschaffel et al. (2009) believe that sociocultural and emotional factors affecting the subject play a leading role in developing and displaying students' strategies for flexibility/adaptability. Of course, in order to illustrate this, several surveys should be carried out in the future.

2.2. Presentation and comparison of two curricula: England and the Netherlands

In this section we present and compare the curricula of mental calculation from two different countries: the programme of England, the National Numeracy Strategy (NNS), and the Realistic Mathematics Education (RME) of the Netherlands. These two curricula are among the most modern ones; additionally, many studies and comparisons have been conducted throughout the international literature (Beishuizen & Anghileri, 1998; Beishuizen, 2001; Anghileri, 2001; Murphy,

2003). For these programmes we will try to identify the key points, the way they have evolved and their philosophy of teaching mental calculation.

2.2.1. Dutch Realistic Mathematics and mental calculation

The Dutch Realistic Mathematics Education (RME) approach has been in effect since the '80s and has introduced important reforms. Regarding mental calculation, and most math content in general, it has characteristics of the modern teaching. These constitute the point of reference and have influenced many educational programmes worldwide. The teaching of mathematics in primary education in Holland over the last thirty years has been greatly influenced by the research of Freudenthal (1973) and his colleagues (Treffers, 1993). The term *Realistic Mathematics Education* (RME) refers to a comprehensive philosophy, which considers mathematics as a 'human activity'. It refers to connections already made between education, mathematics and everyday experience. Mathematics is considered an integral part of real life. Mathematics lessons should give students a 'guided' opportunity to 're-invent' mathematics by doing it. This means that in mathematics education, the focal point should not be mathematics as a closed system, but rather on the activity, on the process of mathematisation (Freudenthal, 1968).

In the early '80s, the first generation of textbooks written according to the theory of RME was published in Holland, based on the first programmes, called 'Wiskobas', which were prevalent the '70s (Treffers, 1991a). In these first RME textbooks, there was an acute perception – as in the British post-Cockcroft period – that discovering number patterns was more important than knowing number facts and that an emphasis on mental strategies will necessarily enhance the learning of numbers without additional training. However, this view seemed not to be confirmed in school reality or research programmes.

During the '80s several comparative studies between traditional and new 'realistic' manuals took place in Holland, which led to heated debates, since the results were not always as expected. One explanation for the poor results of students using RME textbooks was that there was a clear delay in their knowledge of basic arithmetic facts (Gravemeijer, 1994, p. 156).

As Van den Heuvel-Panhuizen (1992) explained, the automation of arithmetic facts, such as supplements of 10, is an important prerequisite for mental calculation with flexibility. If there is not enough practise in retrieval from memory such things as numerical facts, as happened in the first generation of 'realistic' manuals, many children will continue to use strategies of step-by-step counting and will not use more effective cognitive strategies.

After many revision conferences, the Freudenthal Institute published a balanced view of RME in a new curriculum document: *Specimen of a National Program for Primary Mathematics Teaching* (Treffers & De Moor, 1990). This programme remained the expression of the dominant central principles of RME, as the use of problems in context in order to promote students' informal cognitive strategies in problem solving, and interactive discussions in the whole classroom within different strategies. However, more emphasis was put on practising basic arithmetic facts and basic arithmetic skills up to 20 and up to 100.

The empty number line was also introduced to support these two aspects of mental arithmetic: (a) proceduralisation and (b) strategy development (Beishuizen, 1997). We will present these concepts below (see Section 2.2.4.3). These new views on RME led to revisions and the so-called second generation of the 'realistic' school textbooks.

A major difference between the Dutch and the English systems of instruction is the Dutch choice to learn the place value and the written algorithms later, while informal students' cognitive strategies are initially developed and structured. Another difference lies in the Dutch theory of teaching, which is *progressive mathematisation* on which a developmental path or *learning-teaching trajectory* is followed.

The logic of Dutch Realistic Mathematics Education in teaching mental calculation will appear below, in comparison with the teaching process in both countries.

2.2.2. Mental calculations in the numeracy programme in England

A major reform in mathematics education in England took place in 1999 with the National Numeracy Strategy, guided by the *Framework for teaching mathematics from reception to year 6* (DfEE, 1999). One of the four basic principles of the National Numeracy Strategy was an emphasis on mental calculation. The framework contains a set of annual teaching programmes, covering all aspects of the National Curriculum for Mathematics in Primary Education. Networks were also designed to show how mathematical issues in a number of indicative courses can be grouped into work groups for each subject.

Mathematics was proposed to be taught daily by means of the method of a tripartite structure, starting with oral work and mental calculation. The main part of the course was used to teach either new topics or repeat and consolidate previous work. Finally, recapitulation took place, at times where new knowledge was encountered during the teaching.

2.2.3. Common points of the two programmes

To note the main commonalities in the teaching approaches of the two countries, we can first observe that great emphasis is given to mental calculation for teaching mathematics. Apart from this general statement, Murphy (2003, p. 126) notes the following five common points:

- an emphasis is given to the development of mental calculation prior to the teaching of standard written algorithms;
- the recommendation of explicit instruction of these strategies;
- the recognition of the use and value of informal strategies and their place in children's learning;
- a progression from informal strategies to formal strategies;
- guidance on the didactic progression of informal strategies.

Both countries can also be complimented for using the empty number line and giving great importance to the discussion of strategies, whether individually discovered or used by pupils as a whole class.

2.2.4. The differences between the two programmes

Nevertheless, the two educational systems under consideration present significant differences in the development and progress of the suggested instruction, but also in the mode of presentation of the strategies for mental calculation.

2.2.4.1. Introduction and development of strategies

Within the framework of the National Numeracy Strategy (NNS) (DfEE, 1999) in England, an annual programme and its didactic objectives were proposed. In the teaching objectives of each class from Year 1 to Year 6 under the heading 'calculations' there is a reference to strategies of mental calculation and the quick recall of number facts with regard to the four operations. In the introduction of the Framework (DfEE, 1999, p. 6), and through the description of mental strategies in the teaching objectives, an evolutionary process is proposed for teaching mental calculation.

> In the early years, children will use oral methods, in general moving from counting objects or fingers one by one to more sophisticated mental counting strategies. Later they will use a number line or square to work out their answers in different ways, depending on the numbers involved. After giving them experience of a variety of situations, real and imagined, you should teach them to remember and recall simple number facts such as 5 add 3 is 8 or that 7 taken from 9 leaves 2. Posing problems and expressing relationships in different ways, and encouraging children to use this language when they talk about mathematics, is an important stage in developing their calculation strategies and problem-solving skills.

This evolutionary process begins with early counting strategies, then students remember and recall simple number facts and, finally, they use number facts to develop strategies and solve more complex problems (Murphy, 2003).

In the Dutch school of Realistic Mathematics Education, we can clearly see the presentation and development of strategies in mental calculation through the instructional guidance of the project *Tussendoelen Annex Leerlijnen* (TAL), which is addressed to teachers and is an aid to their work (Van den Heuvel-Panhuizen, 2001). This title of the project (TAL), has been translated into English as *Intermediate Attainment Targets in Learning–Teaching Trajectories*. In this work, the development of mental calculation does not occur in isolation, but in conjunction with the development of numbers and in the more general context of calculations with whole numbers. Learning–Teaching Trajectories define the learning procedure that the children will follow. This procedure is continuous and long-term, referring not only to learning, but also to the teaching that needs to be developed and the means to be used. The trajectory should not be seen as a linear step-by-step process, but the provision of the skill levels associated with the calculations of whole numbers.

Concerning mental calculations up to 20, the proposed development is for a *counting* level to a *structuring* level and then at a *formal* level of mental calculation. For mental calculations up to 100, the evolution refers more to the last two levels, of structuring and formal, without excluding the first level. At the *counting* level, operations concentrate on moving along the counting line: jumping forward in addition, jumping back in subtraction. At the *structuring* level, numbers are grouped or decomposed in a handy way. Here structuring materials, such as an arithmetic rank, play a central role. Lastly, on the *formal* level calculations are mainly associated with the relationships of numbers that have already been learned and understood by children.

These three skill levels are associated with three basic types of strategies belonging to calculations up to 100: the *stringing* strategy (N10), the *splitting* strategy (1010) and the *varying* strategy. The *stringing* strategy involves moving along the number line, starting with the first number, and has its roots mainly at the counting level (43 + 28 = 43 + 20 + 7 + 1). The *splitting* strategy involves the 'separation' of the two numbers based on the decimal structure and calculating with the separated parts (43 + 28 = 40 + 20 + 3 + 8). This is somewhat more complex and is mainly connected with the structuring level. Finally, the *varying* strategy is mainly linked to the formal level, which is intrinsic to the logical steps that are done through the relations of numbers and properties of operations (43 + 28 = 43 + 30 − 2). As children become confident in using these three strategies and begin to use them at higher levels, they develop the skills needed to perform calculations up to 100 (Van den Heuvel-Panhuizen, 2001, p. 95).

Murphy (2003, p. 129) constructed a table to show the progression of the strategies of the programme of the National Numeracy Strategy (NNS) in England with examples from Year 1 up to Year 3 (aged six to eight years) and confronted them with the strategies of the Realistic Mathematics Education programme. As noted by Murphy, the selection of the strategies of the English programme is by no means exhaustive. This choice of strategies was to help to compare the development between the two schools.

In Murphy's table, it was shown that strategies belonging to different conceptual levels of Realistic Mathematics were simultaneously introduced in the same year as the programme of England. For example, in Year 2 we have the stringing, splitting and varying strategies belonging to three different conceptual levels: the counting, the structuring and the formal level. Then, by comparing the two ways of progress we can notice a major difference. In England, every year at the same time a number of strategies are introduced and are developed as the complexity of problems grows. The flexibility of students is within the annual objectives; that is they are expected to be able to perform a certain calculation using one or more of the strategies introduced.

In the Dutch school, conceptual levels were proposed – levels in understanding – rather than achievement levels, based on Realistic Mathematics Education research and the theory of 'progressive mathematisation'. It is argued that, because there are a number of different strategies that can be used to solve a problem, the goal of achievement is not only certified by the problems that the students are able to solve. Instead, the understanding levels at which the problems are solved are more important (Van den Heuvel-Panhuizen, 2001, p. 17).

In Dutch Realistic Mathematics Education there are a more limited number of strategies and instructional guidance moves from counting strategies to structuring strategies, which is an extension of the early use of calculation for children. This development is supported by the use of the *empty number line*. Children are guided from a personal, informal level to a more efficient method. The use of productive strategies based on known number facts and the understanding of number is encouraged.

At this point, comparing the two schools, Murphy (2003, pp. 130–131) says:

> In this sense the RME provides a pathway for guided 'reinvention' of calculation strategies where children's personal conceptual levels are recognised. The progression presented in the NNS Framework does not reflect the development through conceptual levels as presented by the Dutch approach. The development of more efficient 'structuring' strategies based on children's early knowledge of counting is less emphasised. The expectation is for children to progress from simple counting strategies to the recall of known facts. A range of deductive strategies that employ these known facts are introduced to solve more sophisticated problems.

Another important difference, which we will examine in detail below, is the simultaneous introduction of stringing strategies (N10) and splitting strategies (1010) in the English programme, which is radically different from the perception of the Dutch programme.

2.2.4.2. Counting and the N10 strategy vs the procedures of place value and the 1010 strategy

In England, under the influence of Piagetian theory and modern mathematics, the place value of the numbers within the context of numbering system has become dominant. Many mental methods as well as standard algorithms of the operations are based on the previous statement. Numbers are separated into HTU (Hundreds–Tens–Units) and students try to find the quantity that each digit in a number represents.

Counting is considered a mechanical and a meaningless process. The use of counting in calculations is considered 'primitive' (Askew & Wiliam, 1995). According to this conception, counting should not be cultivated but instead replaced as soon as possible by the concept of place value and corresponding vertical written procedures for addition and subtraction, which are considered more important. These processes of place value are supported by materials having the structure of a numeric system, such as the abacus, tables of 100, etc.

The strategy of *splitting* (1010) corresponds to this concept of place value and structure of HTU (e.g., $36 + 25 \rightarrow 30 + 20 = 50$ and $6 + 5 = 11$, answer $50 + 11 = 61$). So the strategy (1010) is widely used in England, because of the perception mentioned above.

Freudenthal (1973), as we know, was opposed to the logic of Modern Mathematics and activated the movement called Wiskobas, of Realistic Mathematics Education in Holland. So Freudenthal, against an early emphasis on formal mathematical structures, supported by visual representations and objects, presented the informal and early math skills of the child, meaning counting and structuring strategies. The child initially counts one by one and then discovers the ability to count by two, by five and by ten, which are easy procedures for him. The counting by ones can be extended to counting by tens to numbers up to 100. These procedures of counting forward or backward correspond to the stringing strategy (N10) (e.g., $36 + 25 \rightarrow 36 + 20 = 56, 56 + 5 = 61$). Initially, the use of this strategy is difficult, the counting by ten (e.g., 26, 36, 46, . . . and inversely) is difficult enough for students. With the help of the model of the ENL (Empty Number Line), the N10 strategy is easy to understand and use (Klein et al., 1998).

So the procedures of counting are initially informal and spontaneous, while progressively, with time, they become more effective and constructive activities, as opposed to the procedures of place value by which we separate the numbers based on the standard structure HTU (Hundreds–Tens–Units).

According to the logic of RME, then, in the initial procedure of numerical concepts one of the 'few big ideas' is the reinvention of informal *mental strategies* based on counting, rather than the learning of computational procedures based on place value. This type of guided learning possibly takes a bit more time in the beginning. However, this type of teaching allows students to use their own (procedural) constructions and gradual development of both cognitive and metacognitive strategies (Beishuizen & Anghileri, 1998, p. 523).

So the Dutch RME strategy of stringing (N10) on the number line is introduced first, as a continuation and shortcut to counting. Later, in the second grade, it is followed by the strategy of splitting (1010) as a second basic mental strategy. Then, in the third and fourth grades, these strategies evolve briefly in mixed and diversified methods based on personal preference, while the standard written algorithms are also introduced. During this development, the emphasis moves from specific strategies to more general procedures, which present steps of mental calculation using an informal language of numbers based on efficiency of solutions.

In contrast, in the English programme, various types of mental calculation strategies (1010, N10, etc.) are simultaneously introduced. The introduction of the empty number line ENL is also fragmentary. According to the Dutch perspective and experience, these conditions are not the best to promote new mental strategies (Thompson, 2001).

2.2.4.3. Progressive mathematisation and the empty number line

In Dutch Realistic Mathematics Education, the answer to the question whether mental calculation strategies can be taught or whether pupils spontaneously develop their own methods lies in the middle. Emphasis is placed on students' informal strategies, but their development is not left to chance (Beishuizen & Anghileri, 1998).

Various situations in context are proposed to evoke the use of specific strategies in students. These contextual situations, presented to the students as problems, are consistent with the logic of Realistic Mathematics. For this purpose, several models are designed to encourage students to use counting strategies in the context of everyday life. One such model is the bus model for additions and subtractions up to 20 (Van den Brink, 1991) (e.g., on a bus there are 12 passengers; at a stop, it takes on three passengers and none got off: how many passengers stayed on the bus?). In this way, students are driven by real and familiar situations to situations and strategies more abstract and more mathematically formal. This teaching process of guided development from informal to formal higher level strategies is called *progressive mathematisation* (Treffers, 1991b). It is complete not only through cognitive, but also through metacognitive activation: control of the implementation of the strategy of the written record and reasoning for the choice of the strategy through the discussion of the whole class.

In the Netherlands in the early '90s, the second generation of the realistic school's textbooks introduced the number line and a modified form of it, the empty number line. When used by students, it improved their performance in arithmetic operations and in the development of mental strategies. The number line and the empty number line are suitable materials to help students in the process of strategy internalisation for mental calculation (Murphy, 2008). The number line is suitable to be applied in this procedure of counting, which is directly associated with the operation of addition. In this way, therefore, students applying counting on the number line were gradually led to the implementation of various strategies for operations of calculation. We will then show the way the number line teaching was inserted and showed in the Netherlands, initially for the numbers and then in the operations of addition and subtraction in numbers up to 100.

At first, students received practise in the normal (graduated) number line, and then the empty number line was shown, with the help of which they can perform addition, subtraction, and other operations with natural numbers, including decimals and fractions. For example, in the following figure, students' answers to the problems 35 + 27 and 48 − 26 using ENL are presented.

35 + 27

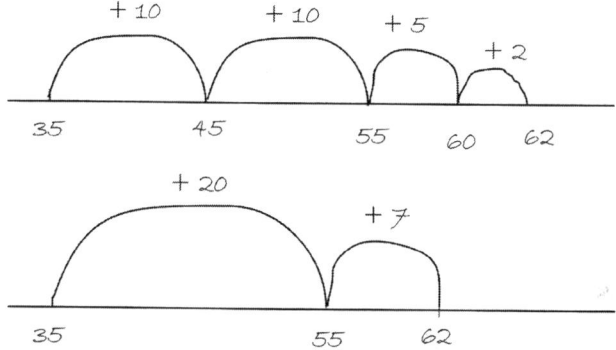

48 − 26

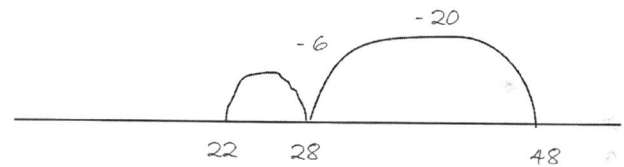

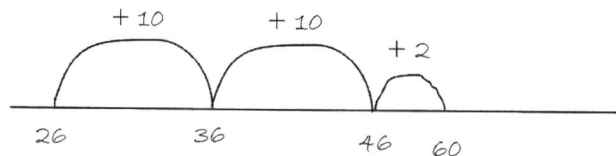

The use of the empty number line is based on the procedure of counting, for example, with the addition 35 + 27, students start from 35 and climb, incrementally, by ten and ten and five and two. As we know, the procedure of counting one by one is familiar to students and is the first informal procedure used to calculate additions and subtractions.

As we saw above, regarding solutions with mental strategies, Treffers (1998) identified three levels in ascending order, namely: (a) review *with counting*, in which, when necessary, it is supported by counting material, (b) *structured* calculation, carried out without counting, but by means of suitable models and (c) *formal* calculation, performed using numbers as mental objects, for short and flexible calculations without using structured materials or models.

Menne (2001, p. 97) states the following:

> For being able to calculate at level two (structured calculating) children should be familiar with some fundamental skills. These skills may be divided into those relating to operations with numbers. Skills concerning number concepts are:
> - counting from any number;
> - locating numbers in relation to each other;
> - identifying jumps towards a number.
>
> Skills concerning number operations are:
> - complements in 10;
> - splitting numbers up to 10;
> - jumps of 10.

Some examples of training in these fundamental skills will be presented below.

Counting on the number line

Students using the number line for operations should be able not only to count on it forwards and backwards, but also to count by one and by jumps with larger numbers as well.

Counting by one starting from a given number

For example, starting from 4 and going up, starting from 8 and descending.

Counting with jumps

Counting by 10: 10, 20, 30, . . . and labelling tens on the number line. This can also be counting by 5 and by 2.

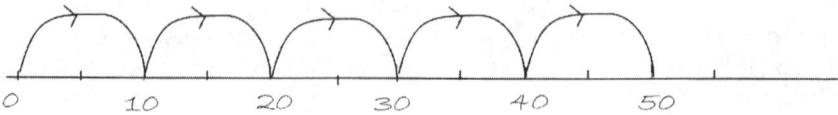

Counting by jumping from a given number

For example, starting from 13 and going up by 10: 13, 23, 33, 43, . . .
Starting from 46 and descending by 10: 46, 36, 26, 16.

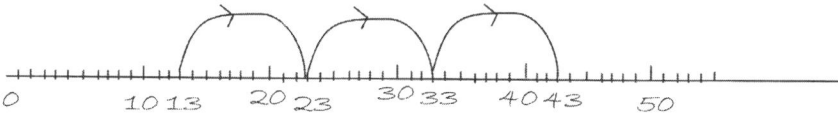

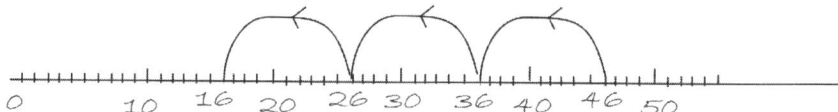

Placing numbers on number line

Placing a number on a number line and determining its position in relation to other numbers is a very important process, later used in mental calculation. For example, in the addition problem 56 + 28, where one calculates as follows: 56 + 30 = 86 → 86 − 2 = 84, one may round 28 at 30, he realises that 28 is close to 30, differing by two.

Number placement activities on the number line

For example, we can have an unfinished number line and ask students to complete some numbers.
   Place on the following number line the numbers: 15, 21, 38 and 44.

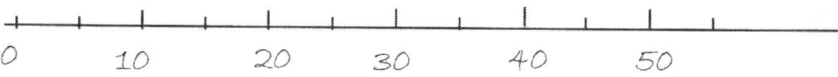

Activities of number placement on imaginary number line

We know that it is important for students to imagine a number line and work on it in their minds. We can play the game 'guess the number'. One student notes a number on his number line with a piece of plasticine. Another student, without looking at it, tries to guess the number by posing questions. The challenge is to try to ask as few questions as possible.

Jumping on the number line

During the teaching of the technique of jumps the structure of numbers is analysed and the relationships between jumps and numbers is discovered. These activities are a natural continuation of the counting process and constitute an essential preparation for arithmetical operations. According to the process of counting, jumps of

10 and hops by one can be made. Thus, the number 18 can be read as a jump of 10 and eight hops.

Students can perform jumps on an imaginary line on the floor, an empty number line etched on the board or on a row of beads. Jumps on an imaginary number line on the floor are important, because children can experiment with the size of the numbers.

'At which number will you jump?' One student makes three jumps forward and two hops backwards. Other students conclude that they reached 28. To check the correctness of their response, they can represent jumps on a row of beads or an empty number line (see below).

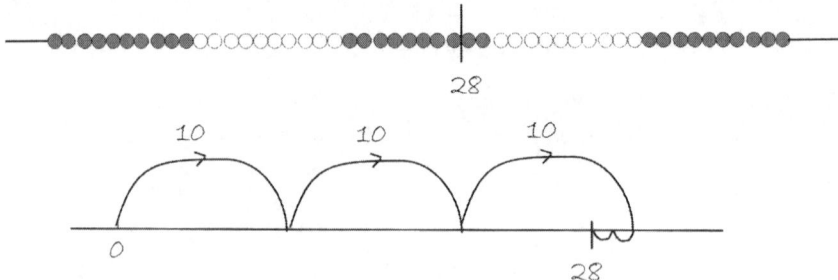

Buys (2001, pp. 108–109) describes the learning and development of mental calculation and their strategies, as shown in the Dutch series of textbooks *Wis en reken* for Year 3 and Year 4 (7- to 9-year-olds) of the Elementary School. This refers to numbers up to 100, but can be extended to numbers up to 1,000. This development, which is presented in the following four stages, can be characterised as 'progressive mathematisation'.

Phase 1: exploring the number domain itself

Various empirical situations are used to make the pupils confident with numbers up to 100. Pupils explore numbers in these situations and get to know their different aspects, such as the structure of the number sequence, the 'environment' of a number (approximate numbers, round numbers, etc.), their position on the number line, the various meanings of numbers in everyday life, etc.; activities on the number line, as described above, belong to this phase.

Phase 2: initiating arithmetic up to 100: jumping in tens and jumping across tens

During the exploration of addition and subtraction problems two basic counting procedures were introduced:

- *Jumping in tens.* For example, 38 + 20 is calculated by performing two jumps of 10: 38, 48 (a jump of 10), 58 (second jump 10).
- *Jumping across tens.* For example, 38 + 6 is calculated by performing two jumps: 38 + 2 = 40, 40 + 4 = 44.

Phase 3: introduction of the sequential method (N10) on the empty number line

First, a general method for solving every problem with numbers up to 100 was introduced. This is the *sequential* method or *stringing* strategy (N10), described in Section 3.3. In this method, the first number is taken as a whole, while the second number is added or subtracted in various chunks. The empty number line is introduced to support this method. Students soon arrive at using this method by just writing the steps used, for example, as shown in the figure below, for the addition of 37 + 28.

Phase 4: extension to two other types of strategies: the split tens (1010) method and varied strategies

When students are adequately familiar with the sequential method, they will become aware successively of two other types of strategies, the *split tens* or *splitting* strategy (1010) and various other strategies. The splitting strategy is used as a second general method to solve any problem with numbers up to 100 (for example, in the addition problem 34 + 23 we have: 30 + 20 = 50, 4 + 3 = 7). The various other methods are based on elementary properties of numbers and operations, such as *compensating* (for example, in addition, we have 37 + 28: 37 + 30 = 67, 67 − 2 = 65), *levelling* (37 + 28: 35 + 30 = 65) and *adding up* (37 to 28: from 28 to 30 is 2, from 30 to 37 is 7, so 2 + 7 = 9).

Two ways of using the empty number line

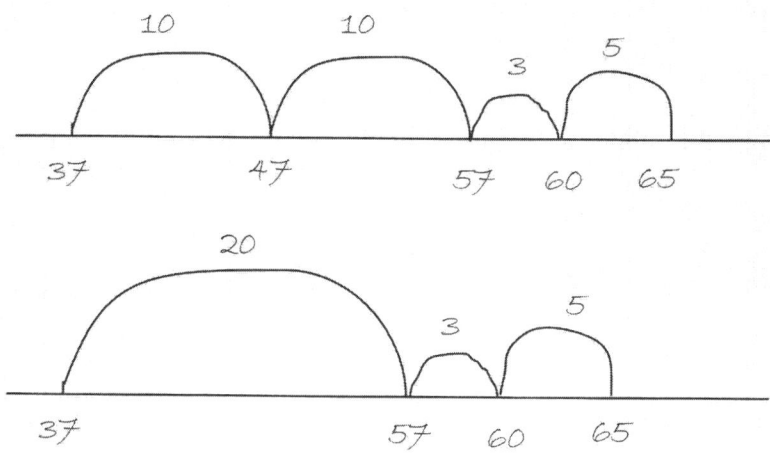

Writing of the operation steps

37+20=57
57+3=60
60+5=65

The use of the empty number line in Dutch Realistic Mathematics influenced the English school. Reading the 'Guidance paper' for the calculations in the new framework for mathematics education (DfES, 2006) in English Primary National Strategy (PNS) it is noted that the use of the empty number line was consistently adopted. Thompson (2007) criticises the use (or misuse) of the empty number line from PNS, saying that it misunderstands the purpose. Murphy (2008) considers that the adoption of the empty number line is fragmented and there are ambiguities. In her investigation Murphy contrasts the presentation of the empty number line (ENL) in the Netherlands with the use of ENL in the teaching of mental calculation in England. To this end, she investigates the adoption by the Dutch the Soviet sociocultural philosophies, especially the Gal'perin theory, and links the concept of internalisation in the analysis of ENL and its use in English Primary National Strategy (PNS).

2.2.4.4. Mental calculations and written formal algorithms

In Dutch Realistic Mathematics, regarding written operations, students at the end of primary education should be able to perform multi-digit addition, subtraction and multiplication with the standard written algorithm (Van den Heuvel-Panhuizen & Wijers, 2005, p. 295). The written operations of multiplication include only single-digit multiplications with multi-digits numbers. The traditional written operation of division is not included at all within the goals of primary education.

Column calculation is an alternative way compared to standard algorithms and is a clearer way of written calculation. Column calculation is the link between mental calculation and calculations with written standard algorithms. As we see in Table 2.1, the splitting strategy of the number into units, tens, hundreds, etc., is applied in the column calculation. These calculations are performed vertically, but from left to right, as opposed to standard algorithms, performed from right to left. In the column calculation the numbers are used as wholes and not as individual digits, as is done with standard algorithms. The intermediate results are calculated mentally.

In Dutch programmes, in the first half of the third grade, column calculation was introduced concerning the addition and subtraction as a continuation of the splitting strategy in mental calculation, later followed by formal written algorithms. Like addition and subtraction, the operation of multiplication extends in time towards the fixed procedures: the column calculation is introduced in the third grade while the standard algorithm follows. In multiplication, column calculation is also introduced along with its extension, the standard algorithm, which derived from mental calculations. As was previously mentioned, the typical algorithm of division is not taught in primary school.

Regarding the English programme, the same evolutionary procedure is followed in order to introduce formal written operations: students first practise mental calculation, then are taught the column calculation, which is an intermediate link between mental calculation and written algorithms, and finally written algorithms are introduced. Written algorithms of addition and subtraction of two-digit

TABLE 2.1 The two types of written arithmetical operation in Dutch Primary Education

Operation	Addition	Subtraction	Multiplication	Division
Column calculation	574 + 342 ――― 800 110 6 ――― 916	916 − 342 ――― 600 30 4 ――― 574	254 × 6 ――― 1200 300 24 ――― 1524	256 ÷ 15 = 17 remainder 1 250 −150 × 10 ――― 106 × 7 −105 ――― 1
Algorithmic calculation	574 + 342 ――― 916	⑧⑪ 916 − 342 ――― 574	254 × 6 ――― 24 300 1200 ――― 1524	or ③② 254 × 6 ――― 1524

or three-digit numbers are introduced at the end of the third grade. Concerning the multiplication algorithm, at the end of the third grade the multiplication of two-digit numbers by a simple-digit number is introduced, at the end of the fourth grade, the multiplication of two-digit numbers by two-digit numbers is introduced and finally, in the fifth grade, the multiplication of three-digit numbers by two-digit numbers is introduced. Unlike the Netherlands, in the English education system, the written algorithm of division is taught and its learning extends from the end of the third or the beginning of the fourth grade until the end of the fifth grade. At the end of the third or the beginning of fourth grade the written algorithm of division of two-digit numbers by a single-digit number is introduced. The expansion of this operation to division of three-digit numbers by single-digit numbers is at the end of the fourth or the beginning of the fifth grade. Lastly, written division of three-digit numbers by double-digit numbers is taught in the fifth grade, which is the last grade of primary school.

References

Ames, C. & Archer, J. (1988). Achievement goals in the classroom: Student learning strategies and motivation processes. *Journal of Educational Psychology*, 80, 260–267.

Anghileri, J. (1999). Issues in teaching multiplication and division. In I. Thompson (Ed.), *Issues in teaching numeracy in primary schools* (pp. 184–194). Buckingham, U.K.: Open University Press.

Anghileri, J. (2001). *Principles and practices in arithmetic teaching*. Buckingham: Open University Press.

Askew, M., Bibby, T. & Brown, M. (1997). *Raising attainment in primary numeracy*. London: King's College, London.

Askew, M. & Wiliam, D. (1995). *Recent research in mathematics education*. London: HMSO.

Australian Education Council (AEC) (1991). *A national statement on mathematics for Australian schools*. Carlton, Vic.: Curriculum Corporation.

Baroody, A. J. (2003). The development of adaptive expertise and flexibility: The integration of conceptual and procedural knowledge. In A. J. Baroody & A. Dowker (Eds.), *The development of arithmetic concepts and skills: Constructing adaptive expertise* (pp. 1–34). Mahwah, NJ: Erlbaum.

Baxter, J. A., Woodward, J. & Olson, D. (2001). Effects of reform-based mathematics instruction on low achievers in five third-grade classrooms. *Elementary School Journal, 101*, 529–547.

Beishuizen, M. (1997). Mental arithmetic: Mental recall or mental strategies? *Mathematics Teaching, 160*, 16–19.

Beishuizen, M. (2001). Different approaches to mastering mental calculation strategies. In J. Anghileri (Ed.), *Principles and practices of arithmetic teaching* (pp. 119–130). Buckingham: Open University Press.

Beishuizen, M. & Anghileri, J. (1998). Which mental strategies in the early number curriculum? A comparison of British ideas and Dutch views. *British Educational Research Journal, 24*(5), 519–538.

Blöte, A. W., Klein, A. S. & Beishuizen, M. (2000). Mental computation and conceptual understanding. *Learning and Instruction, 10*, 221–247.

Blöte, A. W., Van der Burg, E. & Klein, A. S. (2001). Students' flexibility in solving two-digit addition and subtraction problems: Instruction effects. *Journal of Educational Psychology, 93*, 627–638.

Boaler, J. (1998). Open and closed mathematics: Student experiences and understandings. *Journal for Research in Mathematics Education, 29*, 41–62.

Borkowski, J. G. (1985). Signs of intelligence: Strategy generalization and metacognition. In S. R. Yussen (Ed.), *The growth of reflection in children* (pp. 106–144). Orlando, FL: Academic Press.

Borkowski, J. G. & Kurtz, B. E. (1987). Metacognition and executive control. In J. G. Borkowski & J. D. Day (Eds.), *Cognition in special children: Comparative approaches to retardation, learning disabilities, and giftedness* (pp. 123–152). Norwood, NJ: Ablex.

Bottge, B. A., Heinrichs, M., Mehta, Z. & Hung, Y. (2002). Weighing the benefits of anchored math instruction for students with disabilities in general education classes. *The Journal of Special Education, 35*, 186–200.

Bransford, J. D. & Schwartz, D. L. (1999). Rethinking transfer: A simple proposal with multiple implications. *Review of Research in Education, 24*, 61–100.

Buys, K. (2001). Progressive mathematization: Sketch of a learning strand. In J. Anghileri (Ed.), *Principles and practices in arithmetic teaching: Innovative approaches for the primary classroom* (pp. 107–118). Buckingham: Open University Press.

Buzeika, A. (1999). Invented algorithms: Teachers face the challenge. In J. M. Truran & K. M. Truran (Eds.), *Making the difference. Proceedings of the Sixteenth Annual Conference of the Mathematics Education Research Group of Australasia* (pp. 128–134). Sydney: MERGA.

Campione, J. C. & Brown, A. L. (1977). Memory and metamemory development in educable retarded children. In R. V. Kail & J. W. Hagen (Eds.), *Perspectives on the development of memory and cognition* (pp. 367–406). Hillsdale, NJ: Erlbaum.

Carpenter, T. P., Franke, M. L., Jacobs, V. R., Fennema, E. & Empson, S. B. (1998). A longitudinal study of invention and understanding in children's multidigit addition and subtraction. *Journal for Research in Mathematics Education, 29*, 3–20.

Cichon, D. & Ellis, J. G. (2003). The effects of MATH Connections on student achievement, confidence, and perception. In S. L. Senk & D. R. Thompson (Eds.), *Standards-based school mathematics curricula: What are they? What do students learn?* (pp. 345–374). Mahwah, NJ: Lawrence Erlbaum Associates.

Cockcroft, W. H. (1982). *Mathematics counts*. HMSO, London.
Cooper, T., Heirdsfield, A. & Irons, C. (1996). Children's mental strategies for addition and subtraction word problems. In J. Mulligan & M. Mitchelmore (Eds.), *Children's number learning* (pp. 147–162). Adelaide: Australian Association of Mathematics Teachers, Inc.
DES (1989). *Mathematics: The non-statutory guidance*. London: National Curriculum Council.
DfE (1995). *Key Stages 1 and 2 of the National Curriculum*. London: HMSO.
DfEE (1999). *The National Curriculum: Framework for teaching mathematics from reception to year 6*. London: DfEE.
DfES (2006). *Primary national strategy: Primary framework for literacy and mathematics*. London: DfES.
DfES (2007). *Primary Framework for literacy and mathematics: Guidance paper: Calculation*. http://www.standards.dfes.gov.uk/primaryframeworks/mathematics/Papers/Calculation/.
Earl, L., Fullan, M., Leithwood, K. & Watson, N. (2000). *Watching and learning, summary: First annual report*. Toronto: OISE/UT.
Fabricius, W. V. & Hagen, J. W. (1984). Use of causal attributions about recall performance to assess metamemory and predict strategic memory behavior in young children. *Developmental Psychology*, 20, 975–987.
Flavell, J. H. (1970). Developmental studies of mediated memory. In H. W. Reese & L. P. Lipsitt (Eds.), *Advances in child development and behavior* (Vol. 5, pp. 181–211). New York: Academic Press.
Flavell, J. H. & Wellman, H. M. (1977). Metamemory. In R. V. Kail & J. W. Hagen (Eds.), *Perspectives on the development of memory and cognition* (pp. 3–34). Hillsdale, NJ: Erlbaum.
Foxman, D. and Beishuizen, M. (1999). Untaught mental calculation methods used by 11-year-olds. *Mathematics in School*, 28(5), 5–7.
Frensch, P. A. & Sternberg, R. J. (1989). Expertise and intelligent thinking: When is it worse to know better? In R. J. Sternberg (Ed.), *Advances in the psychology of human intelligence* (Vol. 5, pp. 157–188). Hillsdale, NJ: Erlbaum.
Freudenthal, H. (1968). Why to teach mathematics so as to be useful. *Educational Studies in Mathematics*, 1, 3–8.
Freudenthal, H. (1973). *Mathematics as an educational task*. Dordrecht: D. Reidel Publishing.
Fuson, K., Carroll, W. & Drueck, J. (2000). Achievement results for second and third graders using the standards-based curriculum Everyday Mathematics. *Journal for Research in Mathematics Education*, 31, 277–295.
Garcia, T. & Pintrich, P. R. (1994). Regulating motivation and cognition in the classroom: The role of self-schemas and self-regulatory strategies. In D. H. Schunk & B. J. Zimmerman (Eds.), *Self-regulation of learning and performance: Issues and educational applications* (pp. 127–154). Hillsdale, NJ: Erlbaum.
Garner, R. (1990). When children and adults do not use learning strategies: Toward a theory of settings. *Review of Educational Research*, 60, 517–529.
Geary, D. C. (2003). Arithmetical development: Commentary on chapters 9 through 15 and future directions. In A. J. Baroody & A. Dowker (Eds.), *The development of arithmetic concepts and skills: Constructing adaptive expertise* (pp. 453–464). Mahwah, NJ: Erlbaum.
Goos, M., Galbraith, P. & Renshaw, P. (1999). Establishing a community of practice in a secondary mathematics classroom. In L. Burton (Ed.), *Learning mathematics: From hierarchies to networks* (pp. 51–74). London: Falmer Press.
Gravemeijer, K. (1994) *Developing realistic mathematics education* (PhD dissertation). Utrecht, Freudenthal Institute.
Gravemeijer, K. (2004). Local instruction theories as means of support for teachers in reform mathematics education. *Mathematical Thinking and Learning*, 6, 105–128.
Gray, E. (1991). An analysis of diverging approaches to simple arithmetic: Preference and its consequences. *Educational Studies in Mathematics*, 22, 551–574.

Gray, E. (1997). Developing a flexible interpretation of symbols. In I. Thompson (Ed.), *Teaching and learning early number* (pp. 63–72). Open University Press, Great Britain.

Hatano, G. & Oura, Y. (2003). Reconceptualizing school learning using insight from expertise research. *Educational Researcher*, *32*(8), 26–29.

Heinze, A., Marshick, F. & Lipowsky, F. (2009). Addition and subtraction of three-digit numbers: Adaptive strategy use and the influence of instruction in German third grade. *ZDM Mathematics Education*, 41, 591–604.

Heirdsfield, A. M. (2000). Mental computation: Is it more than mental architecture? Paper presented at Annual Meeting of the Australian Association for Research in Education, Sydney. Retrieved 12 November 2006 from http://www.aare.edu.au/00pap/hei00259.htm.

Heirdsfield, A. M. (2006). One teacher's role in promoting understanding in mental computation. In H. L. Chick & J. L. Vincent (Eds.), *Proceedings of the 29th International Group for the Psychology of Mathematics Education* (Vol. 3, pp. 113–120). Melbourne: PME.

Heirdsfield, A. M. & Cooper, T. J. (2004a). Factors affecting the process of proficient mental addition and subtraction: Case studies of flexible and inflexible computers. *Journal of Mathematical Behavior*, *23*, 443–463.

Heirdsfield, A. M. & Cooper, T. J. (2004b). Inaccurate mental addition and subtraction: Causes and compensation. *Focus on Learning Problems in Mathematics*, *26*(3), 43–66.

Hiebert, J. & Wearne, D. (1996). Instruction, understanding, and skill in multidigit addition and subtraction. *Cognition and Instruction*, *14*, 251–283.

Kamii, C. (1985). *Young children reinvent arithmetic: Implications of Piaget's theory*. New York: Teachers College Press.

Kamii, C., Lewis, B. & Jones, S. (1991). Reform in primary education: A constructivist view. *Educational Horizons*, *70*, 19–26.

Kamii, C., Lewis, B. & Livingston, S. J. (1993). Primary arithmetic: Children inventing their own procedures. *The Arithmetic Teacher*, *41*(4), 200–2003.

Klein, A. S., Beishuizen, M. & Treffers, A. (1998). The empty number line in Dutch second grades: Realistic versus gradual program design. *Journal for Research in Mathematics Education*, *29*(4), 443–465.

Klein, T. & Beishuizen, M. (1994). Assessment of flexibility in mental arithmetic. In J. E. H. van Luit (Ed.), *Research on learning and instruction in kindergarten and primary school* (pp. 125–152). Doetinchem, the Netherlands: Graviatt Publishing Company.

Kurtz, B. E. & Weinert, F. E. (1989). Metamemory, memory performance, and causal attributions in gifted and average children. *Journal of Experimental Child Psychology*, *48*, 45–61.

Lave, J. (1988). *Cognition in practice*. Cambridge University Press, Cambridge.

Lave, J. & Wenger, E. (1991). *Situated learning: Legitimate peripheral participation*. Cambridge, UK: Cambridge University Press.

Liguras, G. (2012). *The influence of social and psychological factors of 6th grade students in mental calculation flexibility* (unpublished doctoral dissertation). Department of Primary Education, Florina, University of Western Macedonia.

McClain, K., Cobb, P. & Bowers, J. (1998). A contextual investigation of three-digit addition and subtraction. In L. J. Morrow & M. J. Kenney (Eds.), *The teaching and learning of algorithms in school mathematics* (pp. 141–150). Reston: National Council of Teachers of Mathematics.

Meece, J. L., Blumenfeld, P. C. & Hoyle, R. H. (1988). Students' goal orientations and cognitive engagement in classroom activities. *Journal of Educational Psychology*, *80*, 514–523.

Menne, J. (2001). Jumping ahead: an innovative teaching programme. In J. Anghileri (Ed.), *Principles and practices in arithmetic teaching: Innovative approaches for the primary classroom* (pp. 95–106). Buckingham: Open University Press.

Milo, B. & Ruijssenaars, A. J. J. M. (2002). Strategiegebruik van leerlingen in het speciaal basisonderwijs: begeleiden of sturen? [Strategy instruction in special education: Guided or direct instruction?] *Pedagogische Studien, 79*, 117–129.

Moser Opitz, E. (2001). Mathematical knowledge and progress in the mathematical learning of children with special needs in their first year of school. In *MATHE 2000. Selected papers* (pp. 85–88). Dortmund, Germany: University of Dortmund, Department of Mathematics.

Murphy, C. (2003). A theoretical comparison of the teaching of mental calculation strategies in England and the Netherlands. *Research in Mathematics Education, 5*(1), 123–137.

Murphy, C. (2004). How do children come to use a taught mental calculation strategy? *Educational Studies in Mathematics, 56*, 3–18.

Murphy, C. (2008). The use of the empty number line in England and the Netherlands. In O. Figueras, J. Cortina, S. Alatorre, T. Rojano & A. Sepulveda (Eds.), *Proceedings of the Joint Meeting of PME 32*, (Vol. 4, pp. 9–17). Morélia, México: Cinvestav-UMSNH.

National Council of Teachers of Mathematics (1989). *Curriculum and evaluation standards for school mathematics*. Reston, VA: National Council of Teachers of Mathematics.

National Council of Teachers of Mathematics (2000). *Principles and standards for school mathematics*. Reston, VA: National Council of Teachers of Mathematics.

Pintrich, P. R. & De Groot, E. V. (1990). Motivational and self-regulated learning components of classroom academic performance. *Journal of Educational Psychology, 82*, 33–40.

Resnick, L. B., Bill, V. & Lesgold, S. (1992). Developing thinking abilities in arithmetic class. In A. Demetriou, M. Shayer & A. Efklides (Eds.), *Neo-Piagetian theories of cognitive development: Implications and applications for education* (pp. 210–230). London: Routledge.

SCAA (1997). *The Teaching and Assessment of Number at Key Stage 1–3: Discussion Paper no. 10*. London: SCAA.

Selter, C. (1998). Building on children's mathematics. A teaching experiment in grade three. *Educational Studies in Mathematics, 36*, 1–27.

Selter, C. (2001). Addition and subtraction of three-digit numbers: German elementary children's success, methods and strategies. *Educational Studies in Mathematics, 47*, 145–173.

Sfard, A. (1998). On two metaphors for learning and the dangers of choosing just one. *Educational Researcher, 27*(2), 4–13.

Siegler, R. S. & Campbell, J. (1989). Individual differences in children's strategy choices. In P. L. Ackerman, R. J. Sternberg & R. Glaser (Eds.), *Learning and individual differences: Advances in theory and research* (pp. 218–254). New York: Freeman.

Smith, C. (1999). Pencil and paper numeracy. *Mathematics in School, 28*(4), 10–13.

Steffe, L. (1983). Children's algorithms as schemes. *Educational Studies in Mathematics, 14*, 109–125.

Thompson, I. (1999). Getting your head around mental calculation. In I. Thompson (Ed.), *Issues in teaching numeracy in primary schools* (pp. 145–156). Buckingham, UK: Open University Press.

Thompson, I. (2001). Issues for classroom practices in England. In J. Anghileri (Ed.), *Principles and practices in arithmetic teaching: Innovative approaches for the primary classroom* (pp. 68–78). Buckingham: Open University Press.

Thompson, I. (2007). Deconstructing calculation part 1: Addition. *Mathematics Teaching, 202*, 14–15.

Threlfall, J. (2002). Flexible mental calculation. *Educational Studies in Mathematics, 50*, 29–47.

Threlfall, J. (2009). Strategies and flexibility in mental calculations. *ZDM Mathematics Education, 41*(5), 541–555.

Treffers, A. (1991a). Realistic mathematics education in the Netherlands 1980–1990. In L. Streefland (Ed.), *Realistic mathematics education in primary school* (pp. 11–20). Utrecht: Freudenthal Institute.

Treffers, A. (1991b). Didactical background of a mathematics program for primary education. In L. Streefland (Ed.), *Realistic mathematics education in primary school* (pp. 21–56). Utrecht: Freudenthal Institute.

Treffers, A. (1993). Wiskobas and Freudenthal: Realistic mathematics education. *Educational Studies in Mathematics, 25*(1/2), 89–108.

Treffers, A. (1997). *Progressive mathematisation.* Paper presented on the occasion of the tenth anniversary of the 'Mathe 2000 project' at the Institut fur Mathematik-Didaktik, University Dortmund.

Treffers, A. (1998). *Tussendoelen annex leerlijnen: Hele getallen onderbouw basisschool.* [Intermediate goals annex learning/teaching trajectories: Whole number lower grades in primary school.] Utrecht: Freudenthal, SLO and CED.

Treffers, A. & De Moor, E. (1990). *Proeve van een nationaal programma voor het reken-wiskunde-onderwijs op de basisschool. Deel 2: Basisvaardigheden en cijferen.* [Specimen of a national programme for primary mathematics teaching. Part 2: Basic mental skills and written algorithms.] Tilburg, The Netherlands: Zwijsen.

Treffers, A., De Moor, E. & Feijs, E. (1990). *Proeve van een national programma voor het reken/wiksundeonderwijs op de basisschool. Deel 1. Overzicht einddoelen.* [Towards a national curriculum for mathematics education in the elementary school. Part 1. Overview of the goals]. Tilburg, The Netherlands: Zwijsen.

Van den Brink, J. (1991). Realistic arithmetic education for young children. In L. Streefland (Ed.), *Realistic mathematics education in primary school* (pp. 77–92). Utrecht: Freudenthal Institute.

Van den Heuvel-Panhuizen, M. (1992). Test zelf uw toetsideeen [Test your own ideas on test items], *Willent Bartjens, 11,* 25–28.

Van den Heuzel-Panhuizen, M. (Ed.) (2001). *Children learn mathematics: A learning–teaching trajectory with intermediate attainment targets for calculation with whole numbers in primary schools* (Jackie Senior and Language Unlimited). Utrecht: Freudenthal Institute and National Institute for Curriculum Development.

Van den Heuvel-Panhuizen, M. & Wijers, M. M. (2005). Mathematics standards and curricula in the Netherlands. *Zentrallblatt fur Didaktik der Mathematik, 37*(4), 287–307.

Verschaffel, L., Luwel, K., Torbeyns, J. & Van Dooren, W. (2009). Conceptualizing, investigating, and enhancing adaptive expertise in elementary mathematics education. *European Journal of Psychology of Education, 24*(3), 335–359.

Verschaffel, L., Luwel, K., Torbeyns, J. & Van Dooren, W. (2011). Analyzing and developing strategy flexibility in mathematics education. In J. Elen et al. (Eds.), *Links between beliefs and cognitive flexibility: Lessons learned* (pp. 175–197). London: Springer.

Warner, L. B., Davis, G. E., Alcock, L. J. & Coppolo, J. (2002). Flexible mathematical thinking and multiple representations in middle school mathematics. *Mediterranean Journal for Research in Mathematics Education, 1*(2), 37–61.

Wittmann, E. Ch. & Müller, G. N. (1990–1992). *Handbuch Produktiver Rechenübungen. Vols 1 & 2.* [Handbook of productive arithmetic exercises: Volumes 1 & 2.] Düsseldorf und Stuttgart, Germany: Klett Verlag.

Wittmann, E. Ch., & Müller, G. N. (2004). *Das Zahlenbuch.* [The book of numbers. Mathematics.] Düsseldorf und Stuttgart, Germany: Klett Verlag.

Woodward, J., Monroe, K. & Baxter, J. (2001). Enhancing student achievement on performance assessments in mathematics. *Learning Disabilities Quarterly, 24*(Winter), 33–46.

Yackel, E. & Cobb, P. (1996). Classroom sociomathematical norms and intellectual autonomy. *Journal for Research in Mathematics Education, 27,* 458–477.

3
MENTAL CALCULATION
Addition and subtraction

In this chapter, we will follow the usual order of presentation of teaching, in which children are initially taught the first operations of addition and subtraction with single-digit numbers up to 20 and then the operations of addition and subtraction of two-digit numbers up to 100. The first section will present strategies of addition and subtraction with numbers up to 20. Then, in the second section, we will present the results of research on addition and subtraction up to 20. In the third section we will present strategies of addition and subtraction with numbers up to 100 and, finally, in the fourth section, the results of studies in two-digit and multi-digit additions and subtractions.

3.1. Strategies of addition and subtraction with numbers up to 20

Most studies conducted to identify the procedures or strategies that children use to add and subtract single-digit numbers agree that such strategies are developed according to the following three levels (e.g., Carpenter & Moser, 1982; Fusun, 1992):

- **First level. Strategies of using materials or perceptualisation of numbers.** In this first level, children need to perceptualise numbers to make the operations. They use objects or their fingers to construct a direct model of the operation of addition or subtraction given. For example, in the addition of 2 + 3, the child shows and counts two fingers one by one, then shows and counts another three fingers, finally counting, one by one, all the fingers from the beginning to find the result. This strategy is called *counting all*. We call the

strategies at this level *strategies with materials* and separate those in which children use their fingers from those using objects to model the operation. These strategies are the first that children use to perform additions or subtractions and we encounter them in kindergarten and the first grades of primary school.
- **Second level. Counting strategies.** Children at this level, in order to calculate addition and subtraction, use the sequence of numbers (number line) in contrast to the previous level, in which they counted only objects. For this reason these strategies are called *counting strategies*. For example, in the operation of $2 + 5$, children may count one by one as many steps as the numbers of the operation indicate, starting from the first number $1, 2, \ldots, 3, 4, 5, 6, 7$ (*Counting from the first term*). There are also other strategies of counting detailed below.
- **Third level. Derived fact or construction strategies.** At this level, children retrieve from memory known number facts and treat them mentally to calculate another unknown fact. For example, the addition of $5 + 6$ can be calculated by some children as follows: $5 + 5 + 1 = 11$, retrieving from their memory known number facts: $6 = 5 + 1$ and $5 + 5 = 10$. At this level we distinguish two subcategories of strategies: Strategies of *direct retrieval*, in which the child knows the result of an operation, for example $3 + 3$, by rote. He knows, i.e., the operation and its result and retrieves it immediately from long-term memory. We have also *construction strategies* or *production of operations*, in which the child, in order to find the result of an operation, retrieves from memory known number facts and constructs the answer using these.

This distinction of the three levels is not absolute. There may be strategies that combine behaviours from two different levels; for example, calculation procedures with fingers, where the child computes the result (third level), but confirms it using his fingers (second level).

Hereafter, we will present in tables (Tables 3.1, 3.2 and 3.3) all addition and subtraction strategies and will analyse them further.

3.1.1. First level: strategies using materials or perceptualisation of numbers

3.1.1.1. Addition

Counting all: is one of the first strategies that children use for addition; they often use it spontaneously. In this strategy children need to perceptualise the two terms of addition with fingers or objects.

For example, in the problem, '*Paul has three cards. Mark has two more cards than Paul. How many cards has Mark?*' the student counts using fingers or objects and constructs first a set of 3 objects and a second of 2. Then the two sets are counted together to give the answer (5).

TABLE 3.1 Strategies of addition and subtraction for representing numbers using materials (1st level)

Strategy	Example
1st level: Strategies of using materials or perceptualisation of numbers	
Addition	
1. *Counting all*	**Operation: 2 + 3.** The child counts two fingers. The child counts three fingers. The child counts all the fingers from the beginning.
Subtraction	
1. *Take-away a*	**Operation: 5 − 3**. The child counts five fingers. The child counts and lowers three of them. He counts the remaining fingers.
2. *Separate-to a*	He counts five cubes. The child takes away cubes to get a remainder of three. The child counts what he took in order to find the result.
3. *Add-on-up-to s*	He counts and forms a collection of three cubes. To the three cubes the child adds one by one until they become five. The child counts what is added to find the result.
4. *Match*	He counts and puts in a row three cubes. Underneath the child makes another set of five cubes, which correspond to the above three. The child counts those left to find the answer.

3.1.1.2. Subtraction

Take-away a: here the child constructs the minuend using objects or fingers, then separates the smaller term, the subtrahend, from them, then counting the remainder to give the answer.

We will examine how the various strategies of subtraction are applied in the following problem: 'Paul has 3 cards. Mark has 5 cards. How many more cards does Mark have than Paul?'

Here, the child, using objects or fingers, counts and makes a set that corresponds to the larger number of the problem, the minuend (5), then counts and removes as many objects as the lowest number (3) shows. At the end he counts the remaining objects (2), the result of the subtraction.

Separate-to a: in this problem, the child makes a set that corresponds to the larger number (5) and then removes objects until there remain as many as the lower number shown (3). The answer is found by counting the removed objects (2).

Add-on-up-to s: in this strategy, the child starts from the subtrahend and goes up one by one until he reaches the minuend. The steps that he goes up are the difference. The child, i.e., counts and forms a number of objects (3) equal to the smaller of the two given numbers (the subtrahend). Then adds objects to that set one by one, until he reaches a collection equal to the minuend, the bigger of the given numbers (5). Counting the number of objects added, he has the answer (2).

Match: this strategy is performed only when there are objects. It is impossible to be performed mentally. According to this strategy, the elements of two sets of objects of the terms of subtraction are matched one by one. Counting what is remaining from this matching, we have the answer.

3.1.2. Second level: counting strategies
3.1.2.1. Addition

For the counting strategies for addition we can distinguish two major categories of strategies: the *counting all* and *counting on strategies*. For the *counting-all* strategy the two terms of the addition are counted one by one, e.g., in the addition 3 + 5, a student can answer: 1, 2, 3, . . . , 4, 5, 6, 7, 8. For the *counting on* strategy a restriction in the counting steps is made, so that a student, in the above addition, responds (3), 4, 5, 6, 7, 8. He starts, i.e., from the cardinal of the first term and moves up as many steps as the second term shows. We can say that the *counting on* strategy is more advanced than the *counting all* strategy.

For the *counting on* strategy, we can distinguish two sub-categories: the *count on from the first* strategy and the *count on from the larger* strategy. In the *count on from the larger* strategy, when the second term of addition is bigger than the first, e.g., 2 + 7, we have an informal application of the commutative property of addition, so there is an economy of steps.

Below, in Table 3.2 we present in detail all the strategies of addition and subtraction. For the addition 2 + 7, the following strategies can be applied:

Count all from the first: The first number (2) is counted starting from 1, '1, 2', and counting continues with all terms of the second number (7), '3, 4, 5, 6, 7, 8, 9'. The answer is the last number (9) counted.

Count all from the larger: The commutative property is applied informally, that is, rather than calculate 2 + 7, 7 + 2 is calculated. All the steps of two terms are counted. Counting proceeds up to the larger number (7) starting from 1, '1, 2, 3, 4, 5, 6, 7', and this straight counting is continued by the count of the small number (2) '8, 9'. The answer is the last number (9) counted.

Count on from the first: We have here an economy of steps: children count starting from the cardinality of the first addend. In our example, the child will say 'two (pause)' and count '3, 4, 5, 6, 7, 8, 9'. The answer is 9.

Count on from the larger: As in the previous strategy, here there is also an economy of steps: children start to count from the cardinality of the large addend. By doing so, the commutative property is applied informally. In this example the child will count '7 (pause), 8, 9'. The answer is 9.

3.1.2.2. Subtraction

For the subtraction of 7 − 3 the following strategies can be applied:

Count back from: in this strategy children count back starting from the larger of the two terms of the subtraction. The steps of the count back are as many as the smallest term. The last number pronounced in this back counting is the answer.

TABLE 3.2 Counting strategies of addition and subtraction (2nd level)

Strategy	Example
2nd level: Counting strategies	
Addition 2 + 7	
1. *Count all from the first*	Children count all the steps of the first and second term: '1, 2, ... , 3, 4, 5, 6, 7, 8, 9'.
2. *Count all from the larger*	Children count all the steps of the two terms starting from the largest: '1, 2, 3, 4, 5, 6, 7, ... , 8, 9'.
3. *Count on from the first*	Children count with the first given number: '(2), ... , 3, 4, 5, 6, 7, 8, 9'.
4. *Count on from the larger*	Children count with the largest given number: '(7), ... , 8, 9'.
Subtraction 7–3	
1. *Count back from*	Children start from the large term and count as many steps as the small term shows: '(7), 6, 5, 4'.
2. *Count back to*	Children start from the large term and count back until they reach the small term. Children count the number of steps to find the answer: '(7), 6, 5, 4, 3'.
3. *Addition*	Children start from the small term and count on until they reach the large term. Children count the number of steps to find the answer: '(3), 4, 5, 6, 7'.

For example, children count back starting from 7 and descend 3 number words, '(7), 6, 5, 4'. The last number word in the counting (4) is the answer.

Count back to: here, counting back is performed by starting from the larger of the two terms until we reach the small term. By counting the steps in this counting, we find the answer.

For example, children count inversely starting from 7 and continue until they reach the small number 3: '(7), 6, 5, 4, 3'. The answer is the number of steps counted (4).

Addition: here, students perform addition instead of subtraction. They start from the smallest term of the subtraction (3) and count straight to reach the large number (7): '(3), 4, 5, 6, 7'. Counting the steps taken in this count, the answer is obtained (4).

3.1.2.3. Use of the fingers

In the above strategies of addition and subtraction during counting, children, in order to stop, must record the number of steps they perform. This recording of steps is done very often using the fingers of the hand. Here, the use of the fingers is different from the previous level strategies (first level). In this case the child does not use the fingers to represent the two collections of objects, but to control the evolution of counting and not overpass the result when it is reached. The fingers are used as a means of recording the steps that were performed on counting. In contrast, in the first level he uses the fingers to represent and perceptualise the terms of addition.

For these strategies, the use of the fingers becomes necessary because, when children count, they simultaneously need a way to count the steps in this counting. For example, in the subtraction of $7 - 3$ using the strategy *count back from*, children should count back from 7 and go down 3 number words: '(7), 6, 5, 4'. That is, while reciting the numbers '6, 5, 4' they must simultaneously count from one to three, in order to know when to stop.

3.1.3. Third level: derived fact or construction strategy

3.1.3.1. Addition

Near-doubles: $7 + 6 = 6 + 6 + 1$. *I said 6 and 6 make 12, 13.* Here students use the sum of doubles (n + n), which are easily calculated. They usually use this strategy when the two numbers differ by one unit. But this strategy can be used in other cases too, e.g., $8 + 5 = 5 + 5 + 3$ or $8 + 6 = 6 + 6 + 2$.

Using five: $6 + 7 = 5 + 1 + 5 + 2$. In this strategy, students analyse each addend based on 5. This strategy is used more often when one of the two terms is 5, e.g., $5 + 8 = 5 + 5 + 3 = 13$.

Bridging through ten: $9 + 7$: $9 + 1 = 10$, $10 + 6 = 16$. *I thought that 9 and 1 make 10, and 10 and 6, 16.* This is one of the most popular strategies in school, but it can be difficult for some students. It is suitable for numbers close to 10, such as 8 and 9. More explications of this strategy are presented later.

Compensation: $9 + 5$: $9 + 1 = 10$, $10 + 5 = 15$, $15 - 1 = 14$. With this strategy, students complete one number, such as 9, until terms become easy to add (such as $10 + 5$) and then subtract what is completed ($10 - 1$). This strategy is certainly not as easy for students.

Levelling: $6 + 8$: $7 + 7 = 14$. This strategy is based on the property of levelling, where what is added to one term is subtracted from the other to have the same result. This property is not easy for students. It is applied to numbers that are close to 10, e.g., $7 + 9 = 6 + 10$, $8 + 4 = 10 + 2$. It can certainly be regarded as a strategy of 'bridging through ten' applied automatically. Very few students apply this strategy spontaneously.

3.1.3.2. Subtraction

Using doubles: $14 - 7 = 7$. *I know it because 7 and 7 make 14.* In this strategy students usually know the inverse addition of doubles ($7 + 7$), which is easy to memorise, and, based on this, they calculate the subtraction. Some students may know by heart the subtraction ($14 - 7 = 7$) and simply use inverse addition to justify their answer.

TABLE 3.3 Derived fact or construction strategies of addition and subtraction (3rd level)

Strategy	Example
3rd level: Derived fact or construction strategy	
Addition	
1. *Near-doubles*	7 + 6 = **13** 7 + 6 = 6 + 6 + 1 or 7 + 7 − 1. Children calculate based on the sum of doubles.
2. *Using five*	**6 + 7** 5 + 1 + 5 + 2 = 10 + 3. Children analyse the addends based on 5.
3. *Bridging through ten*	**9 + 7** 9 + 1 = 10, 10 + 6 = 16. Children add to the large term until they reach 10 and then add the rest of the second term.
4. *Compensation*	**9 + 5** 9 + 1 = 10, 10 + 5 = 15, 15 − 1 = 14. Children complete one term, so that the addition becomes easy, and then subtract this complement from the result.
5. *Levelling*	**6 + 8** 7 + 7 = 14. Children add to one term and subtract from the other the same number to reach a known sum.
Subtraction	
1. *Using doubles*	**14 − 7 = 7** 7 + 7 = 14. Children calculate using inverse addition, which is the sum of doubles (n + n).
2. *Near-doubles*	**9 − 5 = 4** 10 − 5 = 5, 5 − 1 = 4. Children calculate based on the subtraction of doubles (2n − n).
3. *Bridging through ten*	**13 − 7** 13 − 3 = 10, 10 − 4 = 6. Children subtract from the large term until they reach 10 and then subtract the rest from the second term.
4. *Subtraction as the inverse of addition*	**7 − 4 = 3** 4 + 3 = 7. Children use the inverse addition to find the difference.

Near-doubles: (9 − 5: 10 − 5 = 5, 5 − 1 = 4). In this strategy, students calculate based on the subtraction of doubles (2n − n): namely, instead of calculating the difference 9 − 5 they calculate 10 − 5 and subtract 1. Although this strategy is used by children, mainly for addition, there are students who use it for subtraction too.

Bridging through ten: 13 − 7: 13 − 3 = 10, 10 − 4 = 6. This strategy is the same as that of addition. The large term is subtracted from until 10 is reached. Then the rest of the second term is subtracted. This strategy is often used in school.

Subtraction as the inverse of addition: $(7 - 4: 4 + 3 = 7)$. *It is 3, because I know that 4 and 3 makes 7.* Here students use the inverse operation of addition, which they know well, to compute the subtraction.

3.1.4. 'Bridging through ten' strategy

As was mentioned above, the *bridging through ten* strategy is one of the most popular strategies, taught in many countries, such as Greece, from the first grade of primary school. But it is a strategy that admittedly creates difficulties for many students.

Teaching this strategy requires knowledge, care and delicate handling from the teacher, so that there are no misconceptions among the students. We will analyse the processes that make up this strategy and the reasons that it is difficult for some students.

A first point to make is that the *bridging through ten* strategy belongs to the third level, making it a construction strategy. This means that students must have reached a level where they have learned and stored number facts in the long-term memory, such as complements of 10, sums of the form $10 + n$, etc. So we should not rush to teach this strategy. Students should be mature and must have stored in their memory enough number facts that they are able to retrieve in short-term memory and compute with. This level is usually reached towards the end of the first grade.

Let us see in detail what operations are needed and how this strategy of bridging through ten is performed. For example, a student has to perform the addition of $7 + 4$ using the strategy of bridging through ten. What should the student do? Analyse the 4 into sum $4 = 3 + 1$, so that one term can be added to 7 to give a sum of 10. Think of the additions $7 + 3 = 10$ and $10 + 1 = 11$. That is, know and have stored in the long-term memory three number facts, $4 = 3 + 1$, $7 + 3 = 10$ and $10 + 1 = 11$. Retrieve these three number facts in the short-term memory, process them and construct the answer. This process of retrieving from the memory and the simultaneous process of three number facts in the short-term memory is not easy for all students.

For all students initially, but also for many others later, the support of materials or representations is needed in order to perform this demanding intellectual activity. *The question is: which material resources or representations are more appropriate to support students in the strategy of bridging through ten?*

3.1.4.1. The empty number line for the representation of the bridging through ten strategy

With the abacus

The abacus is a tool for the easy material representation of the bridging through ten strategy. The abacus is proposed because it has ten beads in each row and the complement of ten appears easily. The presentation of this strategy on the abacus is as follows: In a row of abacus we form the large number of the sum, for example with the addition of $9 + 4$ we form 9. Then we complete beads from the second

Compute with abacus 9 + 4

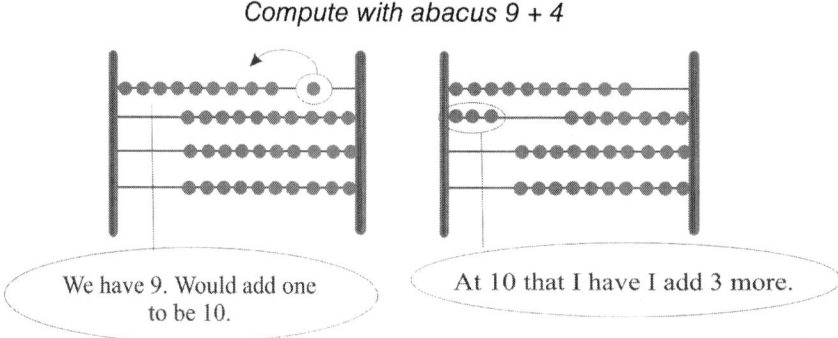

We have 9. Would add one to be 10.

At 10 that I have I add 3 more.

number. We put a bead and complete the first row of abacus. After that we add three more beads in the second row and have the result 13.

With the empty number line

The abacus, however, is not practical and students cannot have it everywhere. It is also not the best representation of numbers and actions performed by the bridging through ten strategy. A handier and more illustrative representation of this strategy is the empty number line. For example, the sum of 7 + 4 in the empty number line is performed as shown in the following figure.

Calculate 7 + 4 with the empty number line

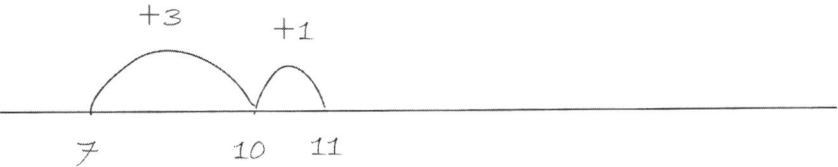

The empty number line is a very good supervisory tool, which can aid students to carry out this strategy. Students monitor the steps of the operation performed and the size and relationship of the numbers on the number line. Moreover, the student, in order to support the implementation of the strategy, can draw a number line very easily whenever he wants.

3.2. Research results for addition and subtraction with numbers up to 20

In this section, we will refer to the results of research on the development of the knowledge of the first strategies of perceptualisation of numbers and counting strategies (first and second level). Also, research showing how children develop retrieval strategies (third level) and how children use these strategies will be presented. Finally, we will refer to the results of studies with Greek students that show the use and development of strategies in the first two years of primary school.

Learning simple addition and subtraction, i.e., sums and differences of one-digit numbers less than ten, is among the most important and crucial learning that the student undertakes in the context of elementary education. It is important because the child will build the knowledge of the other simple operations (multiplication, division) on the basis of knowledge of simple sums and differences. Subsequently, learning written algorithmic operations with multi-digit numbers will be based on these simple operations. Therefore, learning the latter operations is a structural and essential element of knowledge of the four arithmetic operations, as well as of arithmetic in general.

This knowledge is also crucial for the student's attitude towards mathematics. Finding simple sums or differences is among the first problems the child faces in a maths class. The ease of dealing with them and the feeling of success or failure in these first intellectual efforts will be among the determining factors for the development of the attitude of the child toward mathematics, whether positive or negative. For these reasons, we believe that the teaching of simple sums and differences is particularly important and should be treated with due seriousness (Lemonidis, 1998).

Research results, together with teaching experience, show that the speed and accuracy with which children perform every addition and subtraction strategy increases with age and experience. According to this development over time, children not only tend to more frequently use the quickest strategies of the third level (derived fact or construction strategies) but also reduce their use of the slower strategies of the first and second levels, as, e.g., counting all or counting from the first (e.g., Carpenter & Moser, 1982; Geary, Hoard, Byrd-Craven & DeSoto, 2004; Lemonidis, 2008; Siegler, 1987b; Siegler & Shrager, 1984; Starkey & Gelman, 1982).

3.2.1. Development of counting strategies

In the previous section we presented the strategies of addition and subtraction with numbers up to 20, and how they are developed on three levels. Afterwards we will see the results of research on the development of these strategies in children.

3.2.1.1. From the 'counting all or sum' strategy to the 'max and min' strategies

As can be seen in the previous chapter, at an early age and during the early stages of learning, counting is the core competency upon which the learning of arithmetic, and of operations in particular, is based. The number words in the sequence of numbers (one, two, three, . . .), as the numbers, have a double meaning, order (ordinal numbers) and cardinality (cardinal numbers). As pointed out by Fuson and Kwon (1992, p. 291), 'number words to be used in addition and subtraction should acquire the meaning of cardinality'.

In the development of counting as a strategy of addition, some researchers have shown that there are three main stages (Butterworth, 1999, 2005; Carpenter & Moser, 1982):

1. *Counting all* (see first level strategies). For example, for the addition of 2 + 4, children will use objects or their fingers to perceptualise the cardinality of the two terms of addition. They will count 'one, two', then 'one, two, three, four' and then count them all together 'one, two, . . . , six' to get a result. In the literature this strategy is also called 'sum'.
2. *Counting from the first* (see second level strategies). Here, the steps of the first addend are not counted and counting starts from it. In our example, children count '3, 4, 5, 6'. The answer is 6. This strategy is also called 'max'.
3. *Counting from the largest* (see second level strategies). The larger of the two addends is selected and they go up as many steps as the smallest addend is. In our example, children say '4, 5, 6'. This is the quickest and most efficient counting strategy in addition. In this strategy the fewest mistakes are made. This strategy is also called 'min'.

These counting procedures are sometimes performed by children with the help of their fingers (counting strategies with fingers) and sometimes without them (verbal counting strategies) (Siegler & Shrager, 1984).

Obviously, the steps mentioned above do not develop in the same way for all children and are not completely separated and sequential, since students can choose and change their strategy depending on the difficulty of the problem. Another important factor in the development of strategies for students is the method of teaching.

Carpenter and Moser (1982) found that in America, among students five to six years old, there is an important transition in the third level in the first six months of school. It seems that children, even at early stages of the development of addition skills, do not need to count the union of two sets. This has also been concluded in an investigation by Starkey and Gelman (1982), in which researchers showed children two sets, but only one at a time, so that there is no possibility of counting all the components. In these conditions, most three-year-old children could solve 2 + 1, while only a few could solve 4 + 2. The five-year-olds could all solve 2 + 1 and 81% could solve 4 + 2. However, only 56% could solve 2 + 4, which shows that some children could not count from the largest and still counted from the first.

As noted above, the frequency of the use of different strategies changes with age and experience. The strategy of *counting all (sum)* is most often used by infants, but most children stop using it from the second and third grade. The strategy of *counting from the largest (min)* is rarely used by infants, occurs very frequently in the first and second grade and gradually becomes less frequent over time. The strategy of *retrieval* increases steadily during the school year until adulthood, but even high school students continue to use the *min* strategy, splitting 10–30% of the problems with single-digit numbers (Geary, 1994; LeFevre, Sadesky & Bisanz, 1996).

The development of procedural skills is partly related to the improvement of children's conceptual understanding of counting and is reflected in a gradual shift from the frequent use of the *counting all* strategy to the *counting from the largest* strategy (Geary, Bow-Thomas & Yao, 1992; Siegler, 1987b).

3.2.1.2. From counting strategies to derived number facts strategies

The use and practise of counting strategies in addition and subtraction lead to the development of basic number facts (Siegler & Shrager, 1984). Once these representations are formed in long-term memory, they will support the use of problem-solving strategies based on memory. The two bases of these strategies are the *direct retrieval* and *construction strategies* (see third level strategies).

With the *direct retrieval* strategy, children give an answer retrieved from long-term memory that exists as a connection occurring between the problem and the result. For example, to the problem 5 + 5, the child immediately responses 'ten'. Direct retrieval is typically used when the two addends are single-digit and indeed, as we shall see below, in basic sums, such as double sums (n + n), in which memorisation is easy and ahead of other sums.

In construction strategies, children construct the response to short-term memory using number facts they know and retrieve from long-term memory. For example, the sum of 5 + 6, is solved by retrieving the sum of 5 + 5 from long-term memory and adding one to it because children also know that 6 = 5 + 1.

The use of retrieval strategies is tempered by a confidence criterion, i.e., the internal standard by which the child estimates its confidence about the correctness of the retrieved response. Children who have a strict criterion indicate only the answers that they are sure are correct, while children with a lenient criterion state that any retrieved answer, right or wrong, is correct (Siegler, 1988).

With the transition from counting strategies to retrieval strategies, based on memory, rapid responses are achieved and the needs of working memory required by counting strategies are reduced (Delaney, Reder, Staszewski & Ritter, 1998; Geary, Bow-Thomas, Liu & Siegler, 1996; Lemaire & Siegler, 1995).

Groen and Parkman (1972), by measuring reaction times to simple additions with first-grade students, first found that one of the first sums stored in memory is the sum of the form n + n (1 + 1, 2 + 2, 3 + 3, . . .), the 'doubles', as they are termed. Additionally, Woods, Resnick and Groen (1975) found that the first differences stored in memory are the doubles n − n, as well as differences in the form of 2n − n (4 − 2, 6 − 3, . . .). In the quick memorisation of sums of the form, n + n language helps. Phrases such as 'two and two are four', 'three and three are six', because the same word is repeated twice, are easily learned by heart, without many times understanding their meaning. For doubles memorisation often precedes understanding.

Different models of mental organisation of number facts have been proposed. Two of the earliest and well-known models, which predicted the strategies used by students, were ones called the *chronometric models* of Groen and Parkman (1972) and Ashcraft (1982). Chronometric models are based on the regularities of the average or the median time for solving various problems. A basic assumption of these models is that we can use the best prediction of the problem-solving time, to deduce to which strategy this time corresponds.

Ashcraft (1982) discovered a different chronometric model, called the *number facts retrieval model*. Ashcraft (1982) found that the size of the smaller addend was the best predictor of solution times for first graders, that the size of the smaller addend and the size of the squared sum were equally good predictors of times for third graders and that the size of the squared sum was the best predictor of time for students over the third grade. This led him to conclude that first graders consistently use the min strategy, fourth graders and older children consistently use retrieval and third graders use sometimes the min strategy and sometimes retrieval.

Afterwards, these chronometric models were disputed by researchers (see, e.g., Siegler, 1987b). Thus, Siegler and Shrager (1984) propose an *associations model*, according to which children learn to associate 3 + 4 with different answers, but the compound with 7 turns out to be the strongest. This model is a simulation of running a computer that has poor initial performance, but with time learns from experience, which consists of patterns that result from the use of strategies and right and wrong answers.

Dehaene and Cohen (1995, 1997) suggest another model, according to which the retrieval of number facts is supported by a system of neural structures that supports vocal and semantic representations. Number facts are stored as specific verbal links, while subtraction and division require more semantic features. For more information on the storage and retrieval of number facts from memory, see Chapter 7 (Section 7.4.).

3.2.2. Research results of the School of Nature and Life Mathematics in single-digit addition and subtraction

We will present below the results on mental calculation in single-digit addition and subtraction for students in the first and second grades of primary school (Lemonidis, 2008). This research was conducted over two school years with two groups of students, who were taught differently. The control group was taught with traditional teaching of the time and was based on the curriculum before the new curriculum (DEPPS, 2003) and books (2006), while the experimental group was taught with a new concept applied to a large part of the new curriculum and the books that have been in use since 2006. In the first grade, 29 students participated in the control group and 31 students in the experimental group, while in the second grade 35 students participated in each group. Students of both groups were tested with seven tests over two consecutive years in the first and second grade. Each student was examined by personal interview and the examination lasted 20–45 minutes.

3.2.2.1. The key features of the experimental instruction of mental calculation

Experimental instruction used with the experimental group for mental calculations had the following four main features:

A) Gradual transition to mental calculations based on students' abilities

In this instruction, the students' ability to calculate mentally was a goal pursued methodically and over time. Students initially, in order to perform operations, had in many cases a need to represent numbers with objects (perceptualised numbers). Instruction aimed to progressively lead students from strategies of calculation with objects to more abstract strategies, namely mental strategies. Instruction took place in a way that students' prior knowledge and each student's special skills were taken into account.

B) Analysis and synthesis of numbers to sums. Calculation strategies with materials

The second key feature of this instruction was the major emphasis given to the analysis and synthesis of numbers to sum. From the beginning of the instruction the numbers were presented and analysed in aggregate form with the help of objects. Sums that were given special attention were:

- Double sums, i.e., of the form n + n, e.g., 2 + 2, 3 + 3, etc.
- Analysis of the numbers based on five at first and then based on ten. For example, 6 is presented as 5 + 1, 7 as 5 + 2, 13 as 10 + 3, etc.

The analysis of numbers to sum based on five and ten is to note the structure of the fingers in humans and the decimal number system.

In the beginning, the teaching of small numbers was performed with materials (bicolour abacus, bases), that put emphasis on the additional analysis of the numbers. Initially, i.e., students used materials or their fingers for such operations, but because these materials favoured additional structure, students did not count the numbers step by step, but had an overall approach based on additive analysis. In this way we are able to say that students can from the beginning perform calculation strategies with materials.

C) Several methods of semantic representation of quantities of numbers and operations

Quantities of numbers and operations were presented through educational material of teaching through different semantic representations, such as images, symbols (dots, lines, etc.) or abstract points (number words, number digits). These different representations of quantity created both teaching situations of varying difficulty and also requiring different calculation strategies and, each time, a different understanding by students. In the early years of elementary school, the presentation of quantities in numbers and operations, for example, with physical objects or virtual representations or symbolic representations, differentiates the child's behaviour and requires a different type of knowledge management.

D) Teaching with metacognitive processes, discussion and presentation of diversity

The teaching procedure followed for the calculation of operations had the following features: students were asked to explain, each time, the way they had calculated (metacognitive process). The various strategies of the children were exhibited and presented in the classroom and a discussion was held. The teachers were aware of strategies that could be used by students.

Traditional teaching, according to which students were taught in the control group, was the teaching that existed in all Greek schools from 1982 to 2006 and, as regards numbers and operations, had the following characteristics: Mental and estimated calculations were taught little or not at all. Numbers with smaller magnitudes were taught: in the first grade numbers up to 20 were taught and in the second grade numbers up to 100, whereas in experimental teaching in the first grade, numbers up to 100 were taught and in the second grade up to 1,000. There was a distinction between pre-arithmetic and arithmetic concepts and the introduction to numbers was done generally very late. Numbers and operations were justified based on the group theory. Finally, with regard to calculation strategies, there was a strong emphasis on counting strategies and no concern about moving to retrieval and construction strategies.

3.2.2.2. Strategies of students in special addition and subtraction operations

Here we will present the strategies used by students of the two groups for special addition and subtraction operations, in May, at the end of first grade. We chose these operations and term them 'special' because these operations, such as double sums (n + n) and sums and differences of 10, are benchmarks in calculation and are widely used to calculate other operations. Secondly, in some of these operations, such as operations of the form $10 + n$ (e.g., $10 + 6$) and $1n - n$ (e.g., $16 - 6$), students' knowledge about two-digit numbers based on the properties of the numbering system appears. These special operations are the following:

- double sum $n + n$ ($9 + 9$). Difference of the form $2 \times n - n$ ($14-7$);
- sum of single digits equal to 10 ($7 + 3$). Difference from 10 ($10 - 4$);
- sum of the form $10 + n$ ($10 + 6$). Difference of the form $1n - n$ ($16 - 6$);
- sum of the form $1n + n$ ($14 + 4$).

According to Table 3.4, we observe that for students in the experimental group, operations where the strategy of direct retrieval dominates are: 9 + 9, 10 + 6, 16 − 6, 7 + 3 and 10 − 4. In the double sum 9 + 9, in the operation of the form 10 + n (10 + 6) and in the operation of the form 1n − n, the vast majority of students used the strategy of direct retrieval from memory. In the operations of the sum and difference from 10 (7 + 3, 10 − 4), about half of the experimental group of students used the strategy of direct retrieval from memory. In these operations, 7 + 3 and 10 − 4, a significant proportion of students (24% and 36.5%, respectively) used the strategy *retrieval of known operations and calculation*. For example, in the operation 10 − 4 students retrieved from their memory the reverse operation 6 + 4 = 10.

For the students of the control group, we found that the percentages of students who used the strategy of direct retrieval were much smaller than the ones in the experimental group.

In conclusion, we can claim that by the end of first grade, the majority of students in the experimental group had stored number facts in memory and retrieved operations of double sums (n + n), operations of the form 10 + n and 1n − n and also the sums and differences of 10. Students in the experimental group presented a much larger repertoire of strategies.

3.3. Strategies of addition and subtraction with numbers from 20 up to 100

Research literature on mathematics education presents many attempts to organise various strategies used by students in addition and subtraction with numbers up to 100 (Beishuizen, 1985, 1993; Blöte, Klein & Beishuizen, 2000; Cooper, Heirdsfield & Irons, 1996; De Corte & Verschaffel, 1987; Fuson, 1992; Reys, Reys, Nohda, Ishida, Yoshikawa & Shimizu, 1991; Thompson & Smith, 1999; Torbeyns, De Smedt, Verschaffel & Ghèsquiere, 2009; Van den Heuvel-Panhuizen, 2001). A synthesis of these strategies is presented in Table 3.5.

3.3.1. Classification of strategies in addition and subtraction with numbers from 20 up to 100

Beishuizen (1985) presents a classification system that describes two main strategies for mental addition and subtraction. The first is the splitting strategy (1010), in which we split both terms of the operation into units and tens and add them separately (e.g., 43 + 26 → (40 + 3) + (20 + 6) → (40 + 20) + (3 + 6) → 60 + 9 = 69). The second is the stringing strategy (N10), in which one term is stabilised and the tens and units of the other term are added successively (e.g., 43 + 26 → 43 + 20 = 63 → 63 + 6 = 69).

Researchers from around the world, such as Fuson (1992) and Cobb (1995) in the United States, Radatz and Schipper (1988) in Germany and Thompson (1994), Deboys and Pitt (1995) in England accept this separation into two types of strategies.

These two types of strategies (1010 and N10) are theoretically and empirically linked with two fundamentally different approaches to arithmetic operations. The

TABLE 3.4 Success rates and strategies for special addition and subtraction

Operation	Group of students	Success	Retrieval strategies		Counting strategies		Strategies with materials	
			Direct retrieval	Operation retrieval	Counting without fingers	Finger counting	Fingers	Objects
9 + 9	Control	**24 (83%)**	12 (50%)	1 (4%)	1 (4%)	7 (29%)		3 (12.5%)
	Experim	**29 (93.5%)**	25 (86%)	3 (10.5%)		1 (3.5%)		
14 − 7	Control	**21 (72.5%)**	3 (14.5%)	3 (14.5%)		9 (43%)		6 (28.5%)
	Experim	**29 (93.5%)**	1 (3.5%)	16 (55%)	4 (14%)	2 (7%)		6 (20.5%)
7 + 3	Control	**28 (96.5%)**	6 (21.5%)	1 (3.5%)	7 (25%)	13 (46.5%)	1 (3.5%)	
	Experim	**29 (93.5%)**	15 (51.5%)	7 (24%)	3 (10.5%)	3 (10.5%)		
10 − 4	Control	**29 (100%)**	7 (24%)	3 (10.5%)	2 (7%)	14 (48.5%)	3 (10.5%)	
	Experim	**30 (97%)**	13 (43.5%)	11 (36.5%)	4 (13.5%)	1 (3.5%)		1 (3.5%)
10 + 6	Control	**29 (100%)**	16 (55%)		3 (10.5%)	7 (24%)	1 (3.5%)	1 (3.5%)
	Experim	**30 (97%)**	27 (90%)		2 (6.5%)			
16 − 6	Control	**24 (83%)**	10 (41.5%)	4 (16.5%)	1 (4%)	4 (16.5%)		5 (21%)
	Experim	**29 (93.5%)**	23 (79.5%)	3 (10.5%)	1 (3.5%)			2 (7%)
14 + 4	Control	**24 (83%)**	2 (8.5%)	4 (16.5%)	7 (29%)	6 (25%)	1 (4%)	4 (16.5%)
	Experim	**28 (90.5%)**	11 (39.5%)	15 (53.5%)		1 (3.5%)		1 (3.5%)

strategy of splitting (1010) refers to the structure of the decimal system (splitting the number in units and tens), and the strategy of stringing (N10) refers to the successive approximation of the number with counting (Fuson, 1992; Lawler, 1990).

Beishuizen, van Putten and van Mulken (1997) expanded the list in order to include the strategy *bridging through 10* (A10), in which the second term of the operation is separated into two parts, so that the addition or the subtraction of the first term with the part of the second term gives a multiple of ten (e.g., 48 + 26 → 48 + 2 = 50 → 50 + 24 = 74).

In a subsequent publication, Klein, Beishuizen and Treffers (1998) present another strategy, called *compensation* (N10C), in which one term of the operation is rounded to the closest multiple of ten and the result is repaired or compensated by the rounding (e.g., 48 + 26 → 50 + 26 = 76 → 76 − 2 = 74).

Yackel (2001) describes solutions based on collections (*collections based* solutions), where both of the two numbers are separated into parts, usually in tens and units (such as 1010), and strategies based on counting or sequence (*counting* or *sequence based*), which start with one number and connect with each other gradually, piece by piece (like N10).

3.3.1.1. Splitting strategy (1010)

In some countries, such as the UK and Greece, the most common strategy in addition is the method of splitting. It is called that because the numbers added or subtracted are split into multiples of ten and units. This strategy is sometimes called the *partitioning method*. In the Netherlands, this method is called the *1010 process*.

> Zoe 34 + 25
>
> > 59 because 30 and 20 make 50 . . . four and five . . . nine . . . nine and 50 makes 59.

Here Zoe uses the splitting strategy. She separates 34 into 30 and 4, 25 into 20 and 5, adds together the two multiples of ten (30 and 20), adds 4 and 5 and then adds together the two sub-totals (50 and 9), to get the correct answer 59.

This strategy can be divided into two subcategories, using the following terms:

Split from right to left (u-1010) or Units–Tens Split

Students start from the right to left and first add the units, then the tens: In the sum 38 + 25, they first find 5 + 8 = 13, then calculate 30 + 20 = 50 and then sum 13 and 50 to find 63.

Split from left to right (1010), or Tens–Units Split

Here, students start from left to right; they first add up the tens and then the units: In the sum 38 + 25, they first find 30 + 20 = 50, then calculate 5 + 8 = 13 and finally sum up 50 and 13 to find 63.

TABLE 3.5 Strategies of addition and subtraction with numbers up to 100

	Strategy		Examples	
			Addition (38 + 25)	Subtraction (63 − 25)
1	Splitting strategy (1010)		38 + 25: 5 + 8 = 13, 30 + 20 = 50, 63	63 − 25: 13 − 5 = 8, 50 − 20 = 30, 38 = 60, 38
2	Stringing strategy (N10)		38 + 25: 38 + 5 = 43, 43 + 20 = 63	63 − 25: 63 − 5 = 58, 58 − 20 = 38 (subtractive) 63 − 25: 25 + 8 = 33, 33 + 30 = 63, 38 (additive)
2.1	Bridging through multiples of 10 (A10)		38 + 25: 38 + 2 = 40, 40 + 23 = 63	63 − 25: 63 − 3 = 60, 60 − 20 = 40, 40 − 2 = 38 (subtractive) 63 − 25: 25 + **5** = 30, 30 + **33** = 63, 38 (additive)
3	Mix strategy of splitting and stringing (10S)		38 + 25: 30 + 20 = 50, 50 + 8 = 58, 58 + 5 = 63	63 − 25: 60 − 20 = 40, 40 + 3 = 43, 43 − 5 = 38
4	Holistic	Compensation (N10C)	38 + 25: 40 + 25 = 65, 65 − 2 = 63	63 − 25: 63 − 30 = 33, 33 + 5 = 38
		Leveling	38 + 25: 40 + 23 = 63	63 − 25: 68 − 30 = 38
5	Subtraction by Addition (SA) Only pour subtraction			63 − 25: 25 + **5** = 30 30 + **30** = 60 60 + **3** = 63 **5 + 30 + 3** = 38
6	Indirect Subtraction (IS) Only pour subtraction			63 − 25: 63 − **3** = 60 60 − **30** = 30 30 − **5** = 25 **3 + 30 + 5** = 38
7	Counting	By units	38 + 25: 38, 39, 40, . . .	63 − 25: 63, 62, 60, . . .
		By tens or other numbers	38 + 25: 38 + 10 = 48, 48 + 10 = 58, 58 + 5 = 63	63 − 25: 63 − 10 = 53, 53 − 10 = 43, 43 − 5 = 38
8	Mental image of pen and paper algorithm		Children think and mentally perform the method of the standard written algorithm	

In subtraction this splitting strategy is less convenient when the subtrahend number has more units than the minuend number, i.e., we need to borrow. For example, in subtraction 43 – 25, we say 13 – 5 = 8, 30 – 20 = 10, hence we obtain 18.

Whereas, in subtraction 47 – 23, where there is no borrowing issue, this strategy is more simply implemented, 40 – 20 = 20, 7 – 3 = 4, 20 + 4 = 24.

3.3.1.2. Stringing Strategy (N10)

This is found less in the education of UK and Greece than the previous strategy. In the Netherlands, however, it is emphasised and taught first. According to this method, we keep the first term stable, we split the second term into units and tens, and we add or subtract units and tens successively from the first term.

> Andrew 63 – 25
>
> I take 5 from the 63 and 58 remain . . . from 58 I take 20 and I have 38 . . . it is 38.

Alternative names for this strategy in the literature are *sequencing* methods, or, in the Netherlands, where it appears with the abbreviation N10, the *jump* method. This strategy is called jump because it can be represented easily – empirical or mentally – on a number line, where you start from a number and move to the answer with a jump on the line, adding or subtracting the appropriate parts of the second number, as does Helen (see figure below).

> Helen 43 – 27
>
> 27 . . . I take 20 from 43 . . . in order to get 23 . . . and I take three from 23 to make 20 . . . and I take another four to get 16.

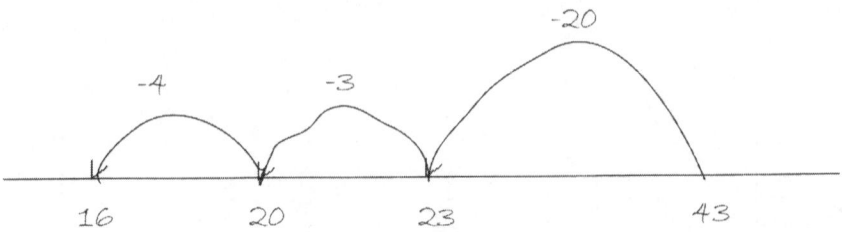

In this strategy, we can distinguish the following two subcategories:

Stringing strategy with units–tens (u-N10)

In the first step we first add or subtract the units and then the tens. For example, in 38 + 25, we add 5 to 38 (38 + 5 = 43), and we add 20 to 43 (43 + 20 = 63).

Stringing strategy with tens–units

To the first term we first add or subtract the tens and then the units. For example, in the subtraction 63 – 25, from 63 we subtract 20 (63 – 20 = 43) and from 43 we subtract 5 (43 – 5 = 38).

3.3.1.2.1 Bridging through multiples of 10 (A10)

This strategy can be considered a subcategory of the stringing strategy (N10) or the independent strategy. The Dutch refer to it with the abbreviation A10. It is a strategy that is used very often in practice, when we calculate mentally. As in the stringing strategy, we keep the first term stable, and add or subtract parts of the second term. The peculiarity is that we add or subtract from the first term such apart from the second term in order to reach the nearest ten, so it is called *bridging through 10*.

> Danae 38 + 25
>
> > I keep 38 and add two to make 40 . . . now from 25 there is 23 left . . . 40 and 23 is 63.

3.3.1.3. Mixed strategy of splitting and stringing (10S)

Many people call it the *split–jump* method, the *mixed* method or, in the Netherlands, method 10S. It is a combination of the splitting and stringing strategy.

> Vangelis 38 + 25
>
> > Sixty-three . . . because, you know, I did 30 and 20 are 50 . . . and 58 . . . and another two from five is 60 . . . and I have three on the right so 63.

In this case, Vangelis splits both numbers: 38 = 30 + 8 and 25 = 20 + 5; he adds 30 + 20 = 50, supplies 8 (the larger of the two units) to 50 (50 + 8 = 58) and adds two more, 58 + 2 = 60, to reach the ten. Finally, he adds the remaining 3: 60 + 3 = 63.

3.3.1.4. Holistic strategies

These strategies are called *holistic* because numbers are used in a holistic way; they are also called *near to multiples of ten*, because they are usually applied to numbers that are close to multiples of ten. This method requires complex cognitive operations, and its use by students is usually an indication of flexibility. A student is flexible when he can select and correctly apply an appropriate strategy, according to the numbers in an operation.

Based on this strategy, students do addition and subtraction in such a way as to have numbers that can be added or subtracted easily. Holistic strategies are divided into two subcategories:

Compensation strategy (N10C)

For this strategy, the abbreviation N10C is used by the Dutch, because it can be regarded as a stringing strategy (N10) that moves to the nearest ten. The compensation strategy is used below by Nick.

> Nick 38 + 25
>
> > Sixty-three . . . 40 and 25 would be 65 . . . if we subtract two we have 63.

We see here that Nick starts with 38 and rounds to the nearest ten, which is 40. So he can easily add 40 + 25 = 65, but in the end he removes the number added to rounding, e.g., two 65 − 2 = 63.

Levelling strategy

This strategy is similar to the previous one, but differs in the following way: what is added or subtracted in the first term of the operation is also added or subtracted in the second term, so that the two interventions cancel out each other in order to have a final result of balance (a + b = c $\Rightarrow$ (a − n) + (b + n) = c).

> Avgi 38 + 25
>
> > Sixty-three . . . I make the 38 40 and the 25 23 . . . so I have 40 and 23 equals 63.

Avgi rounds 38 to the nearest ten, 40, adding 2, but this 2 is removed from 25 and makes 23. So, she adds 40 and 23 and finds 63.

3.3.1.5. Subtraction by addition (SA)

This strategy is only applied to subtractions. Subtraction by addition, as used by Peters, De Smedt, Torbeyns, Ghesquière and Verschaffel (2013), is performed by adding from the subtrahend until reaching to the minuend. Here is an example: 63 − 25 = ?, 25 + 5 = 30, 30 + 30 = 60, 60 + 3 = 63, giving the answer 5 + 30 + 3 = 38. Several authors referred subtraction to addition using different but similar terms. For example, Beishuizen (1997) used the term solving subtractions by means of addition. Selter (2001) used the term the *adding-up* or *completion* strategy. Blöte et al. (2000, 2001) used the term *short jump* strategy, while Torbeyns, De Smedt, Stassens, Ghesquière and Verschaffel (2009) called this strategy *indirect addition*.

3.3.1.6. Indirect subtraction (IS)

This strategy also only applies to subtractions. Indirect subtraction is executed by subtracting from the minuend until getting to the subtrahend, for example, 63 − 25 = ?, 63 − 3 = 60, 60 − 30 = 30, 30 − 5 = 25, so the answer is 3 + 30 + 5 = 38 (De Corte & Verschaffel, 1987, p. 123). This strategy seems to be more effective when there is a small difference between the values in the operation (e.g., 42 − 39 = ?). Previous studies have concluded that the indirect subtraction is used rarely or not at all (Beishuizen et al., 1997; Torbeyns, De Smedt, Stassens, Ghesquière & Verschaffel, 2009).

3.3.1.7. Counting

Here students count step by step based on the sequence of numbers, starting from the first term and going up (addition) or down (subtraction) as many steps as the second term of the operation shows. Counting is a basic method learned in early school age, but when used late it shows weakness in performing mental operations. This strategy is divided into two subcategories, depending on whether students use units or tens in counting:

Counting by units

Starting from the first term, usually the largest, they climb or descend one by one as many steps as the other term shows.

> Anna 38 + 7
>
> Forty-five . . . I said 38 and one is 39 . . . 40, 41, . . . , 45.

Counting by tens

Starting from the first term, they climb or descend by tens or other intervals, according to the way they analyse the other term of the operation.

> Christos 63 – 25
>
> Thirty-eight. . . . I thought 63 . . . I take 10, 53, I take another 10, 43 . . . I take 3, 40, I take 2, 38.

Here Christos starts from the first term, 63, at first goes down by 10 and then by 3 and 2. That is, he analyses the number 25 into 10 + 10 + 3 + 2.

3.3.1.8. Mental image of pen and paper algorithm

Here the method of the standard written algorithm of addition or subtraction is used mentally. That is, students in their minds place the numbers one below the other, as on a sheet of paper, and perform the operation from right to left.

> Alice 38 + 25
>
> Sixty-three . . . I said 8 and 5 make 13 . . . three and one carried . . . one and two, three and three 6 . . . so 63.

3.3.2. Different classifications of strategies

Apart from the categorisation of the strategies outlined above, accepted by a large number of researchers, there are other alternative ways of categorising, the most basic of which are the following.

McIntosh, De Nardi and Swan (1994), based on Skemp (1976), categorised strategies into *instrumental* and *relational*. Instrumental approaches are those showing a particular procedure, without an explanation showing understanding of the concepts involved. In contrast, relational strategies reveal a conceptual understanding of how the numbers work and also show that there is a number sense. This categorisation is used in the following presentation of strategies with rational numbers (see Section 5.1.3).

Lists of strategies are often produced from surveys in which students are presented with computational problems. The strategies used and exposed by students are analysed, so this analysis results in categories. For example, Reys, Reys, Nohda and Emori (1995) use a test in mental calculations to investigate the behaviour and the use of strategies by students in Japan. Before giving the test, the researchers made a detailed categorisation of planned strategies. This categorisation included large clusters and used letters to identify the major strategies and figures for variations of these strategies, e.g., A1, A2, B1, etc. Strategies under the heading A contain calculations in which tens and units are separated (similar to the 1010 strategy), and B includes strategies containing calculations where a number is kept constant (as in strategy N10) and C strategies are those that include rounding one or two numbers to multiples of ten (such as the strategy N10C).

In their research on the effect of gender on the use of strategies by students in first grade, Carr, Jessup and Fuller (1999), categorised the strategies of students into *overt, covert* and *retrieval*. In the category of overt strategies are those strategies that the researcher understands from the external movements of students, such as the use of fingers or objects for mental identification of numbers. In the category covert strategies are strategies where students use their minds, without any external movements. Finally, in the category retrieval, the automatic responses of children are included, where children state that they only know the result by heart.

Van den Heuvel-Panhuizen (2001, p. 123) categorised all strategies for multi-digit addition and subtraction into three main types, based on the way the operation is executed. These categories are *splitting*, *stringing* and *varying*. Splitting strategies (1010) belong in the first type referred to above, while stringing strategies (N10) fit in the second type. Finally, *varying strategies* are holistic strategies (compensation and levelling) which require flexible use of numbers based on the number relations or properties of arithmetic operations.

Torbeyns, De Smedt, Stassens, Ghesquière and Verschaffel (2009) proposed another classification for subtraction strategies based on the operation underlying the solution process. In detail, two main types of strategies are distinguished: (1) direct subtraction (DS), which means the subtrahend is directly subtracted from the minuend and (2) indirect addition (IA), which means adding on to the subtrahend until reaching the minuend. For more information about this classification, see Section 3.4 below.

3.4. Subtraction by addition

This section focuses on studies, published mostly in the last fifteen years, that study solving subtractions by using addition strategies in adults as well as children with and without mathematical learning disabilities (MLD). In Section 2.1.4, the worldwide reform movement of recent decades was highlighted. This movement takes as important goals the flexibility/adaptivity and adaptive expertise of students in mental calculations. Granted that mental subtraction is also an issue in which the variability and flexibility of strategies exist.

In Section 3.1, regarding the subtraction strategies with single-digit numbers up to 20, for the three levels (see Tables 3.1, 3.2 and 3.3), strategies using addition were discussed. For example, Table 3.3 referred to the 'addition' strategy, according to which in subtraction, for example 7 − 3, the student starts from the small term and counts on until they reaching the larger. Then the number of steps to find the answer are counted, for example '(3), 4, 5, 6, 7'.

In the domain of multi-digit subtraction with numbers up to 100, Table 3.5 mentioned the strategy of subtraction by addition (SA) among the different strategies which could be applied. Torbeyns, De Smedt, Stassens, Ghesquière and Verschaffel (2009), as we see below, proposed a classification of subtraction strategies which was based on the operation that underlies the solution process. With that in mind, two main types of strategies are distinguished: (1) *direct subtraction* (DS), which means the subtrahend is directly subtracted from the minuend (e.g., 79 − 42 = ?, 79 − 40 = 39, 39 − 2 = 37) and (2) *indirect addition* (IA), which means adding on to the subtrahend until reaching the minuend (e.g., 79 − 42 = ?, 42 + 30 = 72, 72 + 7 = 79, so the answer is 30 + 7 = 37). This indirect addition strategy we call *subtraction by addition*, following Peters, De Smedt, Torbeyns, Ghesquière and Verschaffel (2013).

There is also a third strategy, (3) called *indirect subtraction* (IS), which involves subtracting from the minuend until you get to the subtrahend (e.g., 79 − 42 = ?, 79 − 30 = 49, 49 − 7 = 42, so the answer is 30 + 7 = 37) (De Corte & Verschaffel, 1987).

Several authors have stated that the subtraction by addition strategy is believed to be especially effective in subtractions (M − S = D) in which a large subtrahend (S) is compared with the difference (D) (e.g., 83 − 79, where S = 79 and D = 4), because this strategy requires fewer and/or smaller calculation steps than the direct subtraction strategy. However, the direct subtraction strategy is considered to be more suitable than subtraction by addition when the subtrahend is smaller than the difference (S < D) (e.g., 83 − 4, where S = 4 and D = 79) (e.g., Anghileri, Beishuizen & van Putten, 2002; Peters, De Smedt, Torbeyns, Ghesquière & Verschaffel, 2013, 2014; Torbeyns, De Smedt, Ghesquière & Verschaffel, 2009a).

Several studies on adults and children were carried out to examine the use of 'subtraction by addition' strategy in multi-digit subtractions relying on verbal protocol data. The results of these studies showed that although adults used the strategy of subtraction by addition efficiently and flexibly (e.g., Peters, De Smedt, Torbeyns, Ghesquière & Verschaffel, 2010a, 2010b; Torbeyns, De Smedt, Peters, Ghesquière & Verschaffel, 2011) children are rarely reported using this strategy (e.g., Blöte et al., 2000; De Smedt, Torbeyns, et al., 2010; Selter, 2001; Torbeyns, De Smedt, Ghesquière & Verschaffel, 2009b).

Peters, De Smedt, Torbeyns, Ghesquière and Verschaffel (2010a, 2010b, 2013, 2014) stated that the method of examination with verbal protocols was not the most appropriate to determine whether children used the indirect addition strategy in multi-digit subtraction. This statement was mainly based upon two observations. First, this strategy was executed by children many times, in a very fast and quasi-automatic way. Secondly, the children seemed to have difficulties expressing precisely the type of strategy they used (see also Ericsson & Simon, 1993; Kirk & Ashcraft, 2001).

Recent studies (Peters, De Smedt, Torbeyns, Ghesquière & Verschaffel, 2012, 2013), using non-verbal research methods, have shown that primary school children actually used the subtraction by addition strategy. Peters, De Smedt, Torbeyns, Ghesquière and Verschaffel used a combination of non-verbal methods to investigate the flexible use of subtraction by addition strategy in two-digit subtraction problems among 72 elementary school children attending the fourth, fifth and sixth grades. The authors found that children of all three grades switched between direct subtraction and subtraction by addition based on the combination of two features of the subtrahend. When the subtrahend was smaller than the difference (e.g., 83 − 4 = ?) the main strategy used was direct subtraction. When the subtrahend was larger than the difference (e.g., 83 − 79 = ?) subtraction by addition was the dominant strategy. The authors concluded that this regularity in the performance of children was observed when there was a large numerical distance between subtrahend and difference.

Still, it would be logical to wonder if variability and flexibility of strategies can be an objective for children with mathematical learning disabilities (MLD). Several researchers and policy makers have claimed that is preferable for these children to possess and use only one way or strategy for the solution of problems (e.g., Geary, 2003; National Mathematics Advisory Panel, 2008). Some studies (e.g., Milo, 2003; Timmermans, 2005) have stated that children with MLD face difficulties in choosing a solution method based on flexibility. However, other researchers supported the idea that flexibility and adaptive expertise can be an objective for children with MLD (e.g., Baroody, 2003; Threlfall, 2002; Peltenburg, van den Heuvel-Panhuizen & Robitzsch, 2012; Peters, De Smedt, Torbeyns, Ghesquière & Verschaffel, 2014; Verschaffel et al., 2007).

Two studies (Peltenburg et al., 2012 and Peters, De Smedt, Torbeyns, Verschaffel & Ghesquière, 2014) examined the use of the subtraction by addition strategy by students with MLD. Peltenburg et al. (2012) found that Dutch special education children (8- to 12-year-olds with a mathematical level of having completed second grade) frequently applied the subtraction by addition strategy spontaneously, although most students had not been taught to use it. Peters, De Smedt, Torbeyns, Verschaffel and Ghesquière (2014) examined 44 children with MLD in Belgium (mean age 12 years and 5 months) using two non-verbal research methods to infer strategy use patterns. They found that children with MLD, like children without MLD, switched between the direct subtraction strategy and the subtraction by addition strategy based on the relative size of the subtrahend. The authors highlighted that these findings were in contrast with typical special education classroom practices, which focused only on the direct subtraction strategy.

3.5. Research results of the School of Nature and Life Mathematics in two-digit and multi-digit additions and subtractions

In this section we will present the results of research conducted on additions and subtractions of two-digit and multi-digit numbers in the School of Nature and Life Mathematics in Greece. Results from worldwide research studies on addition and subtraction with two-digit and multi-digit numbers will not be presented as this was done in Section 2.1., which directly refers to worldwide research results on the teaching of mental computation and flexibility. This issue was discussed to some extent in Section 2.2 too, where there was a comparison of curricula in mental calculation of England and the Netherlands.

3.5.1. Results of a distance learning in-service training programme for elementary school teachers on experimental implementation of mental calculations

In 2009 the laboratory School of Nature and Life Mathematics created an online forum called the Teachers' Network of Nature and Life Mathematics. This Forum was intended to facilitate distance communication with in-service teachers in order to inform and educate them on issues and research results related to the modernisation of mathematics education produced by the scientific team of Nature and Life Mathematics. The first activity developed in this forum was a programme of distance learning for teachers with the experimental application of mental calculation in classrooms.

3.5.1.1. The sample

The programme was implemented in 2010 in all grades of primary school, from first to sixth grade: 63 classes from 25 schools participated from all regions of Greece, totalling 56 teachers and 797 students. The distribution of students among classes and grades was as follows:

TABLE 3.6 Number of classes and students in every grade who participated in the survey

Grades	Number of classes	Number of students
1st	8	88
2nd	12	152
3rd	9	111
4th	11	138
5th	11	138
6th	12	170
Total	**63**	**797**

3.5.1.2. The characteristics of the training programme

The programme was developed in three stages:

1. *Training of teachers and initial evaluation of students.* The training of the teachers, as mentioned above, was implemented through distance learning via the forum. Initially there was discussion between the teachers and researcher on general issues of mental calculation, such as: what is mental calculation? why is it important? what are the differences between mental calculation and written algorithms? and what are the differences and similarities between mental calculation and computational estimation? Meanwhile, different texts for the training of the teachers on the issues of teaching and learning of mental calculation were given. One issue of training given particular emphasis was the strategies that students use in mental calculation. In the context of teacher training in strategies that students use, students of all classes responded to an initial questionnaire on mental calculation. According to this questionnaire, there would be an initial assessment of student ability to perform mental calculation.
2. *Teaching intervention in classrooms.* Each teacher, according to the training, conducted an instruction on mental calculation in the classroom. The main objectives of this were: that students have exercise in mental calculations; that they be driven to discover and implement different strategies in mental operations using natural and rational numbers; and that they exercise expressing oral or written thinking about mental operations. It should be noted that each teacher conducted a free teaching intervention, without fixed schedule or guidance.
3. *Final evaluation of students and evaluation of the programme by teachers.* After the teaching intervention in classrooms, students were tested with the same questionnaire that was used to measure initial knowledge, in order to determine the influence of the teaching intervention on student behaviour. Teachers who participated in the programme were given a questionnaire in order to note benefits gained from their participation in the programme, to describe how they had done their teaching intervention, to examine whether students had improved in relation to the initial questionnaire and to see if there had been a change in the behaviour of their students and, finally, to make free comments and remarks on the programme.

3.5.1.3. Method of examination

As mentioned previously, students were tested twice: by pre-test, before instructional intervention, and by post-test, after intervention. In the test it was noted that the student should calculate mentally and not in a written mode. Also, beside each answer there was a special area, where it was requested for students to note the way in which they were thinking while calculating.

3.5.1.4. Results

We will list below results concerning the behaviours of students in mental additions and subtractions with two-digit and three-digit numbers. These are the success rates before and after the experimental teaching in the second, third, fourth, fifth and sixth grades. The questions in the test for the fifth and sixth grades were the same; several joint questions were also posed to different grades, in order to examine the likely evolution of the performance of students from grade to grade.

3.5.2. Presentation of data from the initial and final questionnaire in second grade

In Tables 3.7, 3.8, 3.9 and 3.10 we present the various grades: the first column shows the correct answers of the students in the classroom, in the next column is the number of students who recorded the strategy they used in mental calculation and then there are the percentages of the various strategies used by students. It should be noted that the rates of strategies are measured for students giving a correct answer. Each column is divided into two columns, where the first presents the results of the pre-test and the second presents the results of the post-test, in bold. In this way we can see the difference in the students' performance created by the teaching intervention.

3.5.2.1. Success in operations and recording capacity of the strategies

In Table 3.7, comparing the success rates in the two sub-columns of the first column, we observe that the success rates for mental calculation do not significantly change before and after teaching. *Hence, teaching intervention in the second grade did not significantly affect the success rates of mental operations.*

Regarding the students' ability to record the calculation strategy that they used for each operation, we observe that, for each operation, a smaller proportion of the students who answered correctly were those who had written down this

TABLE 3.7 Success rates and strategies used by students before and after experimental teaching in second grade

Operations	Correct answers N=152		Number of students recording the strategy		Splitting 1010		Stringing N10		Holistic strategies		Counting		Algorithm	
24 + 25	93	**89.5**	78	**85**	39.5	**46**	2.5	**12.5**	6	**10**	17	**5.5**	13	**11**
58 + 34	85	**81**	68	**79**	35	**49.5**	8	**16**	2	**1.5**	13	**1.5**	10	**11**
26 − 13	78	**83.5**	62	**79**	18	**25**	8.5	**28.5**	4.5	**2**	18.5	**6**	12.5	**18**
49 − 25	73	**80**	62.5	**74**	19	**32**	13	**18**	0.5	**2.5**	15	**4**	12.5	**16.5**

information. For example, in the operation 24 + 25, 93% of the students obtained the correct answer, while only 78% were able to note the strategy used; for the operation 26 − 13 the rates are 78% and 62%, respectively.

Within the objectives of the teaching intervention was also the students' ability to describe verbally or in writing each operation was calculated. We compare in the first and second sub-column of the second column the percentages of students recording their strategy. A statistical comparison of the rates is done by z test comparison rates. *We note that in three of the four operations, a statistical increase in the percentage of students who record the strategy they use is shown.* More specifically, for 58 + 34 we have ($z = 2.2$, $p < 0.05$), for 26 − 13 we have ($z = 3.26$, $p < 0.001$) and for 49 − 25 we have ($z = 2$, $p < 0.05$).

3.5.2.2. Use of strategies

A general observation is that students in the pre-test, before the teaching intervention, predominantly use the splitting strategy (1010): in smaller, but quite significant percentages (ranging from 10% to 18.5%), they use the strategy of counting and the written algorithm; in even smaller proportions (from 2.5% to 13%) the stringing strategy (N10) is used; and finally, very small percentages of students (from 0.5% to 6%) use the holistic strategy. We can say that many students use primitive strategies here, such as counting and the written algorithm. They use holistic strategies less, which shows that they have low flexibility regarding their mental addition and subtraction calculations, although three of the four operations (24 + 25, 58 + 34 and 49 − 25) are offered for holistic strategies.

After the teaching intervention, comparing each time the success rates in both sub-columns, we can observe changes in the use of strategies. The splitting strategy continues to be the one that is used more significantly and with greater rates than in the pre-test (in the pre-test it was used on average in four operations at 28% while in the post-test it was used at 38%). However, the second in usage rates is the stringing strategy (with average use in the four calculations of 18.5%), which shows a very large increase in usage rates compared to the initial state, in which it was the penultimate strategy, with usage rates of an average of 8%. The strategy of counting shows a very large reduction in usage rates (average 5.5%) relative to the initial condition (average 16%). In contrast, the strategy of algorithms is still at high levels, similar to the initial state (initial average 12%, final average 14%). Finally, holistic strategies continue to be used in very small amounts (average 4%) and are the lowest-ranking strategies in usage rates, as in the original condition (average 3.25%).

In conclusion, we can say that teaching intervention in mental calculations changed the behaviour of students in the use of strategies. Students after the teaching intervention continue to use the splitting strategy most; however, they greatly increased the rates of use of the stringing strategy, which they now use as a second strategy. Moreover, several students (averaging 14%) continue to use the strategy

TABLE 3.8 Success rates and strategies used by the students before and after experimental teaching in third grade

Operation	Num. of correct answers N = 111		Number of students recording the strategy		Splitting 1010		Stringing N10		Holistic strategies		Counting	
24 + 25	93.5	98	92	97.5	72	74	5.5	16	1	1	13.5	6.5
79 + 26	80	89	80	89	53	66	14.5	16	3.5	3.5	10	2
58 + 34	85.5	93	85.5	92	61.5	56	11	28	2	2.5	13.5	4.5
147 + 58	63	79.5	68.5	75.5	43	40.5	9	19	5.5	4.5	10	11
49 − 25	79.5	85.5	75	86.5	42.5	40.5	14.5	27	2	4.5	15.5	14.5
80 − 28	55	75.5	64	67.5	29	23.5	23.5	24.5	2	9	10	13.5
71 − 59	38	65	52.5	60.5	28	26	21.5	18	2	2.5	9	12.5
105 − 26	37	51.5	55	51.5	26	23.5	24.5	17	2	2	11	9

of the algorithm; however, they significantly reduced the use of counting strategy (average of 16% to 5.5%). Finally, they continue to use holistic strategies least and with the lowest percentage (4%).

3.5.3. Presentation of data from the initial and final questionnaire in third grade

3.5.3.1. Success in the operations and recording capacity of the strategies

In Table 3.8, comparing the rates in the two sub-columns of correct answers, we observe that there is a statistically significant increase in rates to all questions, except 24 + 25 and 49 − 25, where there were high levels of success before and after the teaching intervention. Hence, for the third grade we could say that with the teaching intervention, we increased the success rates of mental calculation. Regarding the recording capacity of the strategies, we could say that a general change in rates of recording strategies is not shown. Only two of the eight exercises show a statistically significant increase in the percentages of students recording the strategies they use (79 + 26, $z = 1.86$, $p < 0.05$, and 49 − 25, $z = 2.2$, $p < 0.05$).

3.5.3.2. Use of strategies

The splitting strategy is used more than other strategies before and after the teaching intervention, with almost similar usage rates, averaging 44.5% and 43.5%, respectively.

The stringing strategy, before the teaching intervention, seems to be used by students in subtractions more, rather than additions. After the teaching intervention, we have an increased rate of use of the stringing strategy (averaging from 5.5%

to 20.5%), which is statistically significant only in four of the eight calculations: 24 + 25 (z = 2.59, p < 0.01), 58 + 34 (z = 3.22, p < 0.001), 147 + 58 (z = 2.12, p < 0.05), 49 − 25 (z = 2.31, p < 0.05).

The strategy of the written algorithm is used before and after the teaching intervention on average around 10% (11.5% and 9%, respectively). Before the teaching intervention, the strategy of algorithm is used in similar percentages for addition and subtraction. However, after the teaching intervention, it is used in smaller percentages in addition and in almost identical rates in addition with three-digit numbers and in subtraction.

The holistic strategies are those used less before (average 2.5%) and after the teaching intervention (3.5%). This shows that students do not know how to use holistic strategies and they are not at all flexible in the use of strategies before and after teaching intervention.

Finally, in accordance with the above, we can say that in the third grade with the teaching intervention, we have increased the success rates in mental calculations of additions and subtractions. Regarding the use of strategies, we show no significant changes: there is a relative increase in the use of the stringing strategy and a reduction in the use of the written algorithm in two-digit additions, however, students continue to use holistic strategies at lower levels, which means that they remain inflexible in the use of the strategies in addition and subtraction.

3.5.3.3. Comparison of the use of strategies in the second and third grade

In second and third grade there were joint calculations, 24 + 25, 58 + 34 and 49 − 25. According to Tables 3.7 and 3.8, the percentages of correct answers do not generally seem to present major differences in these two grades.

Regarding the recording capacity of the strategies, in the third grade were displayed higher rates of students who could record the strategies they use in relation to second grade, both before and after teaching.

For strategies that students use, in accordance with Tables 3.7 and 3.8, we could say that in third grade students do not use strategies of counting in two-digit addition and subtraction any longer. In third grade, they use splitting strategies in larger percentages than in second grade, both before and after teaching intervention. The other strategies are at similar rates in both grades.

3.5.4. Presentation of data from the pre- and post-test in fourth grade

3.5.4.1. Success in the operations and recording capacity of the strategies

Comparing the success of students before and after instruction, according to Table 3.9, we observe that, after teaching, we have a statistically significant increase in the success rates in the three subtractions that had lower success rates. That is, in

TABLE 3.9 Success rates and strategies used by the students before and after experimental teaching in fourth grade

Operations	Num. of correct answers N = 111		Number of students recording the strategy		Splitting 1010		Stringing N10		Holistic strategies		Algorithm	
79 + 26	89	92	88.5	90	37.5	38.5	21.5	10	23	28.5	5	8.5
58 + 34	89	87	85.5	86	43	40	11.5	14.5	25.5	20.5	5	8
147 + 58	77.5	82.5	79.5	80.5	32	28.5	17.5	16.5	24	25.5	6	10
71 − 59	61.5	73	66.5	70.5	14.5	16	21	26	19	15	9.5	11
80 − 28	71	77.5	75.5	75.5	18	19	30.5	27.5	16	12.5	9.5	16.5
105 − 26	58	71	66.5	69.5	16	12.5	19	19.5	20.5	16.5	6	18
167 − 99	48	60	55	58.5	7	8	22.5	17.5	14.5	20.5	4.5	12.5

71 − 59 (z = 2.05, p < 0.05), in 105 − 26 (z = 2.26, p < 0.05) and in 167 − 99 (z = 2.05, p < 0.05).

Regarding the capacity of recording the strategies, success rates are similar before and after teaching. Hence, teaching intervention in fourth grade did not cause any change in the ability to record strategies.

3.5.4.2. Use of strategies

In fourth grade we have an important change in the use of the strategies in relation to third grade. We are able to compare the use of the strategies in these two grades because six of the seven exercises are common in both grades. Hence, in the fourth grade, the splitting strategy is no more predominant (1010) than it was in second and third grade. In this grade there are used, from now on, all three strategies, namely splitting (prior average 24%, after average 23%), stringing (prior average 20.5%, after average 19%) and holistic strategies (prior average 20.5%, after average 20%), before and after teaching, with percentages almost equally divided, around 20%. Furthermore, for the first time, in fourth grade many students start to use holistic strategies. Finally, we should note that in these three strategies there are no statistically significant differences in the percentages of the use of strategies before and after teaching. Hence, it can be said that the teaching intervention did not influence the use of these strategies.

The use of the algorithm strategy occurs in smaller percentages than the other three strategies. After the teaching intervention, three subtractions 80 − 28 (z = 1.78, p < 0.05), 105 − 26 (z = 3.15, p < 0.001) and 167 − 99 (z = 2.39, p < 0.01), presented statistically significant higher rates of the use of the strategy of the algorithm. Hence, teaching intervention creates an increase in the rate of the use of the strategy of the algorithm on these three subtractions, two of which have a three-digit minuend and lower success rates. It seems here that the students revert to the strategy of the algorithm in difficult subtractions.

TABLE 3.10 Success rates and strategies used by the students before and after experimental teaching in fifth and sixth grade

Operations	Num. of correct answers N = 308		Num. of students recording the strategy		Splitting 1010		Stringing N10		Holistic Strategies		Algorithm	
79 + 26	93.2	96.1	88.6	89	51.3	58.1	7.14	10	10.3	10	20.8	11.4
58 + 34	91	96.1	84.7	89.3	50	56.8	6.2	12.3	7.8	9	19.8	11
147 + 58	80	90	79.2	86.3	44.8	48.7	7.5	11.7	7.5	12.7	18.8	13.3
71 − 59	74	82.8	70.5	77.9	20.1	33.4	14.3	16.2	14.6	14.6	20.1	14
80 − 28	81.8	87.3	75.3	82.5	20.5	28.6	25.6	27.6	11	9	18.5	15.6
105 − 26	71.1	78.2	65.3	72.4	20.5	26.3	12.7	16	12	13	20.1	16.9

3.5.5. Presentation of data from the initial and final questionnaire in fifth and in sixth grade

3.5.5.1. Success in the operation and recording capacity of strategies

According to Table 3.10, if we compare the success rates before and after teaching, in the first and second sub-column of the first column, we can find that after teaching there are statistically significantly higher success rates in all questions are shown. Hence, in fifth and sixth grade after the teaching intervention, we have increased success in calculations of mental addition and subtraction.

Regarding the students' ability to record the strategy that they used, we found that after teaching, except for the first question, all remaining questions have statistically higher rates of recording strategies. Hence, teaching intervention helped the students to note their strategies more frequently.

3.5.5.2. Use of strategies

In fifth and sixth grade, according to the use of strategies, the equipartition that we saw in fourth grade is changing and the splitting strategy is being used more frequently (before teaching process, an average of 34.5% and after an average of 42%), second in frequency is the strategy of the written algorithm (before average 19.5%, after average 14%), third, with lower rates, is the stringing strategy (N10) (before average 12%, after average 15.5%) and, finally, with the lowest rates of use, are the holistic strategies (before average 10.5%, after average 11.5%). Hence, we could say that in the last grades of primary school, children do not use flexible strategies.

We should also note that the highest rates of the use of the splitting strategy (around 50%) appear for the operations of addition, while smaller percentages of this strategy appear in subtraction (around 20%).

The use of the strategy of the written algorithm appears to lower rates after the teaching (before average 19.5%, after average 14%) and the difference in these rates is statistically significant for all additions and the subtraction 71 − 59.

3.5.5.3. Final conclusions and discussion

The above research occurred in correlation with a programme of distance learning for teachers and teaching intervention in all primary school grades in relation to mental calculation. After the training of the teachers in mental calculations and their strategies, each teacher made a teaching intervention in the classroom freely using his own logic and the teaching practice that he wished. As part of this research, we saw the above results from the use of strategies, from the second to the sixth grade in two-digit and three-digit addition and subtraction.

We also evaluated the students' ability to note the strategy they used in mental calculations in the various grades. This metacognitive ability for the students to say or write how they think was significantly emphasised during teacher training. This capacity, recording strategies, was influenced by teaching intervention differently in different grades. In second grade we observed an increase in the percentage of students who were recording their strategies after the teaching intervention. In third grade, regardless of the teaching intervention, the students recorded strategies at higher percentages than second-grade students, and teaching intervention overall did not affect the recording capacity of the strategies, while increasing rates were presented in only two of the eight exercises that were offered to students. In fourth grade, no change was shown in the ability to record after teaching intervention, while the fifth and sixth grade showed an increase in the percentage of students in all questions except one.

Generally, we can observe that there is much room for improvement in students' metacognitive capacity, i.e., the ability to express their thinking in writing concerning mental calculation. This improvement appears after the teaching intervention in some grades like the second, fifth and sixth grades, while in some it is not present, as in the fourth grade. Perhaps this change in the behaviour of students takes time to manifest, and there is a need for a more systematic and long-term intervention.

The strategies that students use for mental calculations in two-digit and three-digit additions and subtractions, as well as the results of the teaching intervention, differ between the various grades of primary school. Regardless, however, of the grades we can make the following two general findings:

First, in all grades of primary school the dominant strategy, namely the one that is used in largest percentages of students, is the splitting strategy (1010). According to this strategy, the terms of the operation are separated in tens and units and are added or removed separately. This splitting strategy resembles a written algorithm. Moreover, every time a sizable percentage of students mentally use the written algorithm. This shows that in Greek schools, the written algorithm of operations dominates in the behaviour of the students.

Secondly, another general observation for all primary school grades is that students do not know how to use holistic strategies; only a significantly smaller percentages of students use these strategies. Furthermore, teaching intervention does not change this behaviour; students still do not use holistic strategies. Only in the fourth grade does the percentage of students who use holistic strategies increase (around 20%), and this occurs despite the teaching intervention. However, this rate is not maintained and decreases below half in upper grades, in fifth and in the sixth grades. A possible explanation of the increase in the percentage of students who use holistic strategies in fourth grade is that in this grade decimal numbers are taught in more depth and perhaps greater importance is given to the digits of numbers and their value, so students see numbers with a holistic perspective. The ability, however, to use holistic strategies that appears in fourth grade is not maintained and is lost in upper grades and students return to more algorithm-oriented methods (increase the use of the written algorithm and splitting strategy). Thus, we could say that students in Greek schools do not know how to use flexible strategies, so they are not flexible in the use of strategies in mental calculations. Nor, moreover, do they use holistic strategies, and they are not flexible even after a teaching intervention.

For the fact that the teaching intervention did not affect the students in the use of holistic strategies we could give the following possible explanations: First, changing the behaviour of students to use more advanced strategies such as holistic, since they are used to operating some well-established strategies, may not be easy: it is possible that teaching intervention needs to be done early, before some strategies consolidate in their behaviour.

Second, teaching intervention was done freely by teachers according to their teaching habits and practices. If we consider, however, the testimony of teachers about how their teaching intervention was done, in addition to the empirical knowledge we have, we can conclude that the strategies were simply presented to the students and taught in a traditional way. So the teaching environment needed to lead students to discover by themselves new strategies, and to be convinced of their effectiveness, is not created in order to increase the use of holistic strategies.

Therefore, if we want a change in the behaviour of Greek students on the issue of flexibility, there is a need to make changes on two fundamental points. First, the teaching of mental calculation in accordance with the logic of flexibility should start from a very early stage. This finding is supported by surveys on teaching flexibility (see Section 2.1.4). Secondly, it is necessary to train teachers on how to teach mental calculation, aiming at the flexibility of students in the use of the strategies.

3.5.6. Research results of flexibility in using strategies of sixth grade students

Ligouras (2012) examined 223 students in sixth grade in Greece, in order to investigate the flexibility of students and the social and psychological factors that influence it. In order to examine the flexibility of students in the use of strategies, he used the research model choice/no choice of Lemaire and Siegler (1995), which is described in Section 1.4.3. Students were proposed 5 addition problems, 5 subtraction problems and 5 multiplication problems, operations which are presented

TABLE 3.11 Proposed operations. From Ligouras (2012), p. 107

A/A	Addition	Subtraction	Multiplication
1	48 + 19	42 − 25	8 × 25
2	39 + 27	71 − 59	9 × 21
3	69 + 56	80 − 28	12 × 18
4	88 + 45	93 − 37	19 × 30
5	147 + 58	167 − 99	15 × 49

in Table 3.11. The numbers in these operations are offered for the use of holistic strategies (e.g., for 48 +19 → 48 + 20 = 68 → 68 − 1 = 67).

According to the method of choice/no choice, the students were tested in personal interviews three times for every operation. At the first time, in the choice mode, students solved each operation freely, using whatever strategy they wished. For the second time, in the no choice mode 1, the students solved the same operations necessarily with holistic strategies and, finally, in the third time, in no choice mode 2, the students solved the operations with any strategy they wanted, apart from holistic strategies.

Ligouras (2012) found that a small percentage of students know and use holistic strategies. Table 3.12 shows the averages of the percentages of students who use holistic strategies for any operation.

Some of the results obtained by the research of Ligouras (2012, pp. 134–135), regarding the use of strategies in additions, subtractions and multiplications are the following:

1. The efficiency and speed in mental calculations are affected by the type of operations (additions, subtractions, multiplications) and the condition under which the operations are performed (free choice, use of only holistic strategies, use of only non-holistic strategies).
2. It is easier for students to add two-digit numbers mentally, and they do this more successfully than to subtract and multiply. The differences in the performance of the students in the three groups of operations are significant.
3. Students find it very difficult to compute with the holistic strategies. Success rates fall greatly when they have to compute only with these strategies. For many children, even when the researcher explained the strategy, it was impossible to calculate using this.

TABLE 3.12 The average rates of use of holistic strategies for each operation. From Ligouras (2012), p. 138

Operation	AVG	STD
Addition	11.03	26.41
Subtraction	15.78	28.13
Multiplication	6.19	18.09
Final	**11.04**	**19.57**

4. Students, when asked to calculate with a strategy different from the holistic, reached success rates similar to those in the choice mode. This means that they know and trust strategies different from the holistic.
5. Two-digit multiplications were very difficult for the students. Success rates were very low, especially in the situation where they had to compute only with holistic strategies.
6. Students calculated more slowly with holistic strategies than any other strategy. This is an unexpected result because these strategies require fewer steps in their implementation. Moreover, previous research (Torbeyns, De Smedt, Ghesquière & Verschaffel, 2009a, 2009b) indicates that the use of holistic strategies require less runtime compared to other strategies. This, however, does not seem to be taken into account by the students who prefer to use a different strategy (1010, N10 or written algorithm) that they know well.
7. The students calculate mentally faster in addition than in subtraction and multiplication in every situation. Only when they had to use a different strategy from holistic was subtraction executed more quickly than addition.
8. It has become evident that the students do not know nor trust holistic strategies, although they offer clear advantages regarding the steps of execution. Thus, they employ other strategies, even in cases where the operations are suitable for the use of holistic strategies, as in the present study. This obviously affects the performance and the speed of execution of operations.

In summary, the results indicate that students are not aware of holistic strategies and find significant difficulties in their utilisation. They know and trust mainly the non-holistic strategies; this became transparent from results showing that there are no significant differences in the performance and the speed between the situation of free choice and the use of non-holistic strategies. This means that in the school there is not a systematic teaching of strategies, nor a discussion about the pros and cons of each strategy. This relates to the curriculum too, which, though it refers to mental calculations, does not specify or arrange their teaching.

References

Anghileri, J., Beishuizen, M. & van Putten, K. (2002). From informal strategies to structured procedures: Mind the gap! *Educational Studies in Mathematics*, *49*(2), 149–170.

Ashcraft, M. H. (1982). The development of mental arithmetic: A chronometric approach. *Developmental Review*, *2*, 213–236.

Baroody, A. J. (2003). The development of adaptive expertise and flexibility: The integration of conceptual and procedural knowledge. In A. J. Baroody & A. Dowker (Eds.), *The development of arithmetic concepts and skills: Constructing adaptive expertise* (pp. 1–33). Mahwah, NJ: Lawrence Erlbaum Associates.

Beishuizen, M. (1985). Evaluation of the use of structured materials in the teaching of primary mathematics. In B. S. Alloway & G. M. Mills (Eds.), *New Directions in education and training technology: Volume 18, Aspects of educational technology* (pp. 246–258). London: Kogan Page.

Beishuizen, M. (1993). Mental strategies and materials or models for addition and subtraction up to 100 in Dutch second grades. *Journal for Research in Mathematics Education, 24*(4), 294–323.

Beishuizen, M. (1997). Development of mathematical strategies and procedures up to 100. In M. Beishuizen, K. P. E. Gravemeijer & E. C. D. M. van Lieshout (Eds.), *The role of contexts and models in the development of mathematical strategies and procedures* (pp. 127–161). Utrecht, The Netherlands: CDβ Press.

Beishuizen, M., van Putten, C. M. & van Mulken, F. (1997). Mental arithmetic and strategy use with indirect number problems up to one hundred. *Learning and Instruction, 7*, 87–106.

Blöte, A. W., Klein, A. S. & Beishuizen, M. (2000). Mental computation and conceptual understanding. *Learning and Instruction, 10*, 221–247.

Blöte, A. W., Van der Burg, E. & Klein, A. S. (2001). Students' flexibility in solving two-digit addition and subtraction problems: Instruction effects. *Journal of Educational Psychology, 93*, 627–638.

Butterworth, B. (1999). *The mathematical brain*. London: Macmillan.

Butterworth, B. (2005). The development of arithmetical abilities. *Journal of Child Psychology and Psychiatry, 46*, 3–18.

Carr, M., Jessup, D. L. & Fuller, D. (1999). Gender differences in first-grade mathematics strategy use: Parent and teacher contributions. *Journal for Research in Mathematics Education, 30*(1).

Carpenter, T. P. & Moser, J. M. (1982). The development of addition and subtraction problem solving skills. In T. P. Carpenter, J. M. Moser & T. A. Romberg (Eds.), *Addition and subtraction: A cognitive perspective* (pp. 9–24). Hillsdale, NJ: Lawrence Erlbaum Associates.

C.C.F. (Δ.Ε.Π.Π.Σ) (2003). *Cross Curriculum Framework*. Pedagogical Institute, Ministry of National Education and Religious Affairs, Government Gazette 303B/13-3-2003.

Cobb, P. (1995). Cultural tools and mathematics learning: A case study. *Journal for Research in Mathematics Education, 26*, 362–385.

Cooper, T. J., Heirdsfield, A. & Irons, C. J. (1996). Children's mental strategies for addition and subtraction word problems. In J. Mulligan & M. Mitchelmore (Eds.), *Children's number learning* (pp. 147–162). Adelaide: Australian Association of Mathematics Teachers, Inc.

Deboys, M. & Pitt, E. (1995). *Lines of development in primary mathematics. Open University set book*. Belfast: Blackstaff Press.

De Corte, E. & Verschaffel, L. (1987). The effect of semantic structure on first graders' strategies for solving addition and subtraction word problems. *Journal for Research in Mathematics Education, 18*, 363–381.

Dehaene, S. (1997). *The number sense: How the mind creates mathematics*. New York: Oxford University Press.

Dehaene, S. & Cohen, L. (1995). Towards an anatomical and functional model of number processing. *Mathematical Cognition, 1*, 83–120.

Delaney, P. F., Reder, L. M., Staszewski, J. J. & Ritter, F. E. (1998). The strategy-specific nature of improvement: The power law applies by strategy within task. *Psychological Science, 9*, 1–7.

De Smedt, B., Torbeyns, J., Stassens, N., Ghesquière, P. & Verschaffel, L. (2010). Frequency, efficiency and flexibility of indirect addition in two learning environments. *Learning and Instruction, 20*, 205–215.

De Smedt, B., Verschaffel, L. & Ghesquiere, P. (2009). The predictive value of numerical magnitude comparison for individual differences in mathematics achievement. *Journal of Experimental Child Psychology, 103*, 469–479. http://dx.doi.org/10.1016/j.jecp.2009.01.010.

Ericsson, K. A. & Simon, H. A. (1993). *Protocol analysis: Verbal reports as data* (Rev. ed.). Cambridge, MA: Bradford/MIT Press.

Fuson, K. (1992). Research on whole number addition and subtraction. In D. A. Grouws, *Handbook of research on mathematics teaching and learning* (pp. 243–275). New York: MacMillan.

Fuson, K. C. & Kwon, Y. (1992). Learning addition and subtraction: Effects of number words and other cultural tools. In J. Bideaud, C. Meljac & J. P. Fisher (Eds.), *Pathways to number, children's developing numerical abilities* (pp. 351–374). Hillsdale, NJ: LEA.

Geary, D. C. (1994). *Children's mathematical development: Research and practical implications*. Washington, DC: American Psychological Association.

Geary, D. C. (2003). Arithmetical development: Commentary on Chapters 9 through 15 and future directions. In A. J. Baroody & A. Dowker (Eds.), *The development of arithmetic concepts and skills: Constructing adaptive expertise* (pp. 453–464). Mahwah, NJ: Lawrence Erlbaum Associates.

Geary, D. C., Bow-Thomas, C. C. & Yao, Y. (1992). Counting knowledge and skill in cognitive addition: A comparison of normal and mathematically disabled children. *Journal of Experimental Child Psychology, 54*, 372–391.

Geary, D. C., Bow-Thomas, C. C., Liu, F. & Siegler, R. S. (1996). Development of arithmetical competencies in Chinese and American children: Influence of age, language, and schooling. *Child Development, 67*, 2022–2044.

Geary, D., Hoard, M., Byrd-Craven, J. & DeSoto, C. (2004). Strategy choices in simple and complex addition: Contributions of working memory and counting knowledge for children with mathematical disability. *Journal of Experimental Child Psychology, 88*(2004), 121–151.

Groen, J. & Parkman, J. M. (1972). A chronometric analysis of simple addition. *Psychological Review, 79*(4), 329–343.

Kirk, E. P. & Ashcraft, M. H. (2001). Telling stories: The perils and promise of using verbal reports to study math strategies. *Journal of Experimental Psychology: Learning, Memory, and Cognition, 27*, 157–175.

Klein, A. S., Beishuizen, M. & Treffers, A. (1998). The empty number line in Dutch second grades: Realistic versus gradual program design. *Journal for Research in Mathematics Education, 29*, 443–464.

Lawler, R. W. (1990). Constructing knowledge from interactions. In L. P. Steffe & T. Wood (Eds.), *Transforming children's mathematics education* (pp. 47–61). Hillsdale, NJ: Erlbaum.

LeFevre, J. A., Sadesky, G. S. & Bisanz, J. (1996). Selection of procedures in mental addition: Reassessing the problem-size effect in adults. *Journal of Experimental Psychology: Learning, Memory, and Cognition, 22*, 216–230.

Lemaire, P. & Siegler, R. S. (1995). Four aspects of strategic change: Contributions to children's learning of multiplication. *Journal of Experimental Psychology: General, 124*, 83–97.

Lemonidis, Ch. (1998). Διδασκαλία των πρώτων αριθμητικών εννοιών. [Teaching of the first arithmetical concepts.] *Ερευνητική διάσταση της Διδακτικής των Μαθηματικών, 3*, 87–122.

Lemonidis, Ch. (2008). Longitudinal study on mental calculation development in the first two grades of primary school. *International Journal for Mathematics in Education. Hellenic Mathematical Society, Volume 1*, 47–68.

Liguras, G. (2012). *The influence of social and psychological factors of sixth grade students in mental calculation flexibility* (unpublished doctoral dissertation). University of Western Macedonia, Florina, Greece.

McIntosh, A. J., De Nardi, E. & Swan, P. (1994). *Think mathematically*. Melbourne: Longman.

Milo, B. F. (2003). *Mathematics instruction for special-needs students. Effects of instructional variants in addition and subtraction up to 100* (unpublished doctoral dissertation). Leiden University, Leiden, The Netherlands.

National Mathematics Advisory Panel (2008). *Foundations for success: The final report of the National Mathematics Advisory Panel*. Washington: U.S. Department of Education.

Peltenburg, M., van den Heuvel-Panhuizen, M. & Robitzsch, A. (2012). Special education students' use of indirect addition in solving subtraction problems up to 100—A proof of the didactical potential of an ignored procedure. *Educational Studies in Mathematics*, *79*(3), 351–369.

Peters, G., De Smedt, B., Torbeyns, J., Ghesquière, P. & Verschaffel, L. (2010a). Using addition to solve large subtractions in the number domain up to 20. *Acta Psychologica*, *133*, 163–169.

Peters, G., De Smedt, B., Torbeyns, J., Ghesquière, P. & Verschaffel, L. (2010b). Adults' use of subtraction by addition. *Acta Psychologica*, *135*, 323–329.

Peters, G., De Smedt, B., Torbeyns, J., Ghesquière, P. & Verschaffel, L. (2012). Children's use of subtraction by addition on large single-digit subtractions. *Educational Studies in Mathematics*, *79*, 335–349.

Peters, G., De Smedt, B., Torbeyns, J., Ghesquière, P. & Verschaffel, L. (2013). Children's use of addition to solve two-digit subtraction problems. *British Journal of Psychology*, *104*(4), 495–511.

Peters, G., De Smedt, B., Torbeyns, J., Ghesquière, P. & Verschaffel, L. (2014). Using addition to solve subtraction problems in the number domain up to 20 and 100. *MENON: Journal Of Educational Research. 1st Thematic Issue*, 8–27.

Peters, G., De Smedt, B., Torbeyns, J., Verschaffel, L. & Ghesquière, P. (2014). Subtraction by addition in children with mathematical learning disabilities. *Learning and Instruction*, *30*, 1–8.

Radatz, H. & Schipper, W. (1988). *Handbuch für den Mathematikunterricht an Grundschulen.* [Handbook for mathematics education in primary school.] Hannover: Schroedel Schulbuch Verlang.

Reys, R. E., Reys, B., Nohda, N., Ishida, J., Yoshikawa, S. & Shimizu, K. (1991). Computational estimation performance and strategies used by fifth- and eighth-grade Japanese students. *Journal for Research in Mathematics Education*, *22*(1), 39–58.

Reys, R. E., Reys, B. J., Nohda, N. & Emori, H. (1995). Mental computation performance and strategy use of Japanese students in grades 2, 4, 6, and 8. *Journal for Research in Mathematics Education*, *26*(4), 304–326.

Selter, C. (2001). Addition and subtraction of three-digit numbers: German elementary children's success, methods and strategies. *Educational Studies in Mathematics*, *47*, 145–173.

Siegler, R. S. (1987). The perils of averaging data over strategies: An example from children's addition. *Journal of Experimental Psychology: General*, *116*, 250–264.

Siegler, R. S. (1988). Individual differences in strategy choices: Good students, not-so-good students, and perfectionists. *Child Development*, 59, 833–851.

Siegler, R. S. & Shrager, J. (1984). Strategy choices in addition and subtraction: How do children know what to do? In C. Sophian (Ed.), *The origins of cognitive skills* (pp. 229–293). Hillsdale, NJ: Erlbaum.

Skemp, R. R. (1976). Relational understanding and instrumental understanding. *Mathematics Teaching*, 77, 20–26.

Starkey, P. & Gelman, R. (1982). The development of addition and subtraction abilities prior to formal schooling in arithmetic. In T. P. Carpenter, J. M. Moser & T. A. Romberg (Eds.), Addition and subtraction: A cognitive perspective (pp. 99–116). Hillsdale, NJ: LEA.

Thompson, I. (1994). Young children's idiosyncratic written algorithms for addition. *Educational Studies in Mathematics*, *26*, 323–345.

Thompson, I. & Smith, F. (1999). *Mental calculation strategies for the addition and subtraction of 2-digit numbers. Final report.* University of Newcastle, Newcastle upon Tyne.

Threlfall, J. (2002). Flexible mental calculation. *Educational Studies in Mathematics*, *50*, 29–47.

Timmermans, R. E. (2005). *Addition and subtraction strategies: Assessment and instruction*. Nijmegen: Radboud Universiteit Nijmegen.

Torbeyns, J., De Smedt, B., Ghesquière, P. & Verschaffel, L. (2009a). Solving subtractions adaptively by means of indirect addition: Influence of task, subject, and instructional factors. *Mediterranean Journal for Research in Mathematics Education*, *8*(2), 1–30.

Torbeyns, J., De Smedt, B., Ghesquière, P. & Verschaffel, L. (2009b). Acquisition and use of shortcut strategies by traditionally schooled children. *Educational Studies in Mathematics*, *71*, 1–17.

Torbeyns, J., De Smedt, B., Stassens, N., Ghesquiere, P. & Verschaffel, L. (2009). Solving subtraction problems by means of indirect addition. *Mathematical Thinking and Learning*, *11*, 79–91.

Torbeyns, J., De Smedt, B., Peters, G., Ghesquière, P. & Verschaffel L. (2011). Use of indirect addition in adults' mental subtraction in the number domain up to 1,000. *British Journal of Psychology*, *102*, 585–597.

Van den Heuvel-Panhuizen, M. (Ed.) (2001). *Children learn mathematics*. Utrecht: Freudenthal Institute, Utrecht University.

Verschaffel, L., Torbeyns, J., De Smedt, B., Luwel, K. & Van Dooren, W. (2007). Strategy flexibility in children with low achievement in mathematics. *Educational & Child Psychology*, *24*, 16–27.

Woods, S. S., Resnick, L. B. & Groen, G. J. (1975). An experimental test of five process models for subtraction. *Journal of Educational Psychology*, *67*(1), 17–21.

Yackel, E. (2001). Perspectives on arithmetic from classroom based research in the United States of America. In J. Anghileri (Ed.), *Principles and practices in arithmetic teaching* (pp. 15–32). Buckingham: Open University Press.

4
MENTAL CALCULATION
Multiplication and division

This chapter develops the theme of mental calculation with multiplication and divisions. First discussed is how students' knowledge of the multiplication tables and the strategies they use for it develop, together with the correspondent operations of division.

The strategies that students use when performing multi-digit multiplications are analysed and presented. Research results on the study of multi-digit multiplication and division of the School of Nature and Life Mathematics are also presented.

Finally, although this book does not deal with written algorithms, it develops the theme of Greek multiplication. Greek multiplication, what it is and how it is used in teaching are presented in order to introduce the classic written multiplication algorithm.

4.1. Learning the multiplication tables

Research on learning single-digit multiplication, or *multiplication tables*, extends to a large number of issues. Sherin and Fuson (2005, pp. 348–350) classify these studies into four groups based on their dominant characteristics, which are: (a) *semantic types* (models of situations), (b) *intuitive models*, (c) *solution procedures* (computational strategies) and (d) *models of retrieval*.

Research that deals with *semantic types* is interested in the classification of situations described in word problems, according to the way they are formulated before their solution. That is, the content of the word problems lead to different structures, with regard to the mathematical content (Greer, 1992; Kouba, 1989; Marshall, 1995; Nesher, 1988; Reed, 1999; Vergnaud, 1988). These types of situations are: (a) *isomorphism of measures*, (b) *multiplicative factor or measure space*, (c) *product of measures or Cartesian product* and (d) *multiple proportion*.

Intuitive models, which are also called *indirect*, *implicit* or *informal* models, are the research subject of Fischbein, Deri, Nello and Marino (1985), in which the authors state that each fundamental numerical operation remains associated with an indirect, unconscious and primitive intuitive model. The solution of a problem is influenced by this model. In the case of multiplication, this intuitive model is considered to be 'repeated addition'.

The third research issue of multiplication tables refers to the solution procedures. The analysis of *computational strategies* considers the description of the processes sequence executed by a student in order to go from given numbers to reach their product (Kouba, 1989; Mulligan & Mitchelmore, 1997). Finally, the fourth research issue focuses on the nature and development of the models of *retrieval*. Typically, one way of teaching the multiplication tables has been to help students to develop the ability to quickly declare the product of two numbers. Some researchers have dealt with the construction of cognitive models of this ability and how it is developed (Baroody, 1999; LeFevre & Liu, 1997; Lemaire & Siegler, 1995; Siegler, 1988, referenced in Sherin & Fuson, 2005).

These research fields are of course interrelated, and many studies refer to issues that cover two or more of the above groups.

4.1.1. Memory mechanisms and learning multiplication tables

The ultimate goal of teaching the multiplication table in elementary school is for the student to know the products of the multiplication table and to be able to immediately retrieve them from his memory or to calculate them using other learned products. The role of memory in the process of learning the multiplication table is very important, since the acquisition of the calculation skill includes, among other things, memorising basic arithmetic facts.

Multiplication products are stored in an associated memory network, in which activation spreads from the central node to the adjacent nodes. Recent cognitive models (Ashcraft, 1995, in Galfano, Russconi & Umilta, 2003), indicate that, at least for those processes whose factors are represented by single digits, adults retrieve solutions from the stored representations of knowledge, known generally as *arithmetic facts*.

According to Galfano et al. (2003), multiplication seems to be more concerned with retrieval than for the adoption of a procedural strategy, since single-digit multiplication is learned directly with the assistance of memory strategies. Also, several case studies have concluded that the multiplication facts are accessed easier than other simple numerical procedures (Dagenbach & McCloskey, 1992; Girelli, Delazer, Semenza & Denes, 1996; Whetstone, 1998, cited in Galfano et al., 2003). This confirms the existence of a privileged connection between multiplication and retrieval strategies. This link helps explain the poor performance usually observed in multiplication, compared with addition and subtraction, since procedural strategies such as counting do not constitute an important means of assistance in the execution of multiplication (Cohen & Dehaene, 1994, cited in Galfano et al., 2003).

Acquiring ability in arithmetic includes, as mentioned above, among other things, memorising basic arithmetic facts. However, the form in which the arithmetic facts are stored in memory is a matter of controversy. One team of researchers argues that the arithmetic facts are stored on an acoustic or phonological form (Cohen & Dehaene, 2000, cited in Whalen et al., 2002). A form of phonological storage hypothesis suggests that arithmetic facts such as 6 × 8 = 48, are stored in memory phonologically (e.g., six eight, forty-eight). This assumption implies that the arithmetic facts are accessed from memory using a phonological representation of the problem. The answer can then be extracted from the retrieval of arithmetic facts and be converted into the appropriate format for the production of the result (Whalen et al., 2002).

Poor sources of working memory result in a deficient representation of arithmetic facts in long-term memory (Geary, 1990; Siegler & Shrager, 1984, cited in Steel & Funnel, 2001). Working memory is likely to be associated with the selection of appropriate strategies and solving more complex problems (Bull, Johnston & Roy, 1999; Logie, Gilhooly & Wynn, 1994, cited in Steel & Funnel, 2001). Problems containing bigger factors lead to slower responses and more errors than those containing smaller numbers, showing that calculations including larger factors have greater demands on working memory (Hitch, 1978a, 1978b, cited in Steel & Funnel, 2001).

Additionally, neuropsychological studies have shown that arithmetic facts can be selectively lost after a brain injury (Cohen & Dehaene, 1994; McCloskey, Aliminosa & Sokol, 1991, cited in Galfano et al., 2003). Such results can be explained by the fact that such arithmetic facts are stored independently from other conceptual knowledge or other mathematical skills. Therefore, by learning the multiplication tables, arithmetic facts must be stored in memory as shape events that are linked together and not as unrelated events, since without connection they are likely to be forgotten or confused.

4.1.2. Driving mechanisms for strategy development in multiplication tables

Several researchers have investigated mechanisms leading to the development of multiplication strategies. Some argue that mechanisms relating to single-digit multiplication are comparable to those of single-digit addition. Specifically, according to Anghileri (1989), the transition process from *unitary counting* to *rhythmic counting* in multiplication is associated with the transition from the *counting all* process to the *counting on* process in addition. According to her, the transition from one stage to the next is characterised by the child's ability to recognise that the last word during the count of the first set of objects represents the number of its elements.

In contrast, some researchers have linked strategy development in single-digit multiplication with conceptual changes quite different from those described in the previous approach. For example, the research of Mulligan and Mitchelmore (1997) is based on the concept of intuitive models of Fischbein et al. (1985), as an internal mental structure that children apply in situations of multiplication. These researchers studied the developmental progress of 70 girls from the second to the third grade in primary school, and in their results categorised this development into three intuitive models, *direct counting*, *repeated addition* and *multiplicative operation*, and linked each of them with one or more calculation strategies.

1. Direct counting: Unitary counting
2. Repeated addition: Rhythmic counting forwards. Skip counting forwards. Repeated addition. Additive doubling.
3. Multiplicative operation: Known multiplicative fact. Derived multiplicative fact.

- *Direct counting*. The objects are simply counted without any apparent reference to the multiplicative structure.
- *Rhythmic counting*. Counting follows the problem's structure. Simultaneously with counting it takes place a second counting of the number of the groups.
- *Skip counting*. Counting is done with multiples.
- *Repeated addition*. Counting is replaced by calculations such as '3 + 3 = 6, 6 + 3 = 9'.
- *Additive doubling*. For example, '4 and 4 is 8, 8 and 8 is 16'.
- *Multiplicative operation*. The calculation is made by retrieving from memory a known multiplicative operation or by producing it from a known operation. Mulligan and Mitchelmore (1997, p. 316).

Therefore, strategy development in addition and the subtraction is guided by conceptual skills associated with number. Conversely, Sherin and Fuson (2005) believe that during the teaching of multiplication, the most significant changes are not driven by the way in which students perceive quantity. On the contrary, these changes are driven by significant changes in *computational resources in specific numbers*. Students gain a large amount of knowledge about specific numbers – such as 4, 12 and 32 – and this knowledge allows the use of new strategies in a new context. For this reason, many of key issues associated with learning single-digit multiplications are very different from those associated with addition. Experience shows that many students enter into the formal instruction of multiplication already possessing conceptual structures of addition. Moreover, during formal instruction, students already have the basic conceptual skills required for understanding the multiplication. Indeed, it is even argued that in kindergarten students can solve simple multiplication problems (Carpenter, Ansell, Franke, Fennema & Weisbeck, 1993).

There are also significant differences in computational resources used by students in addition and subtraction with respect to multiplication. As shown below, *pattern* learning (rules of 0, 1, 10 and 9) is particularly important in multiplication and can constitute a specific objective in teaching. Also important are computational resources with specific numbers, referred to below by the name *count by sequence*. When learning the multiplication tables, or even before, students learn to count going up by 2 or by 5 (5, 10, 15, 20). Moreover, as in simple addition, they learn arithmetic facts (e.g., 4 × 4 = 16). This is very important for multiplication, because without the knowledge of multiplication facts, calculations can be time-consuming and burdensome for students. Conversely, if a student cannot retrieve a specific addition arithmetic fact, he can often reconstruct the calculation in relatively little time.

4.1.3. Strategies for calculating with multiplication tables

There are several studies in which an attempt is made to organise and present the strategies used in the multiplication tables (Anghileri, 1989; Cooney, Swanson & Ladd, 1988; Kouba, 1989; LeFevre, Bisanz, Daley, Buffone, Greenham & Sadesky, 1996; Lemaire & Siegler, 1995; Mulligan & Mitchelmore, 1997; Sherin & Fuson, 2005; Siegler, 1988). In these studies there is no consensus on the terminology used for strategies, nor for the number and variability of different strategies. Some studies examine students solving word problems (Anghileri, 1989; Kouba, 1989; Mulligan & Mitchelmore, 1997), while others, referred to above, use only arithmetic operations. Additionally, in some of the studies students were allowed to use manipulative objects or pencil and paper (e.g., Anghileri, 1989), while others did not allow any external help. Further, in the above studies the age of the students tested was different. For example, in Anghileri's study (1989), students 4–12 years old were examined, in Kouba's (1989) there were students from the first, second and third grade of primary school, while in LeFevre et al.'s (1996) study, adults were examined.

Research by Sherin and Fuson (2005), attempts a comprehensive and systematic classification of strategies in the multiplication tables. In this research, on the one hand, the results of the previous literature on strategies in the multiplication tables were used and analysed, and, on the other hand, results are derived from long-term experimental application with third- and fourth-grade students within the programme Children's Math Worlds (CMW).

In this research, therefore, strategy groups are related to the type of specific computational resources that support these strategies. That is, a group of strategies associated with a specific pattern of use of one or more types of specific computational resources are described. There are six main groups of strategies presented: (1) *count all*, (2) *additive calculation*, (3) *count by*, (4) *pattern based*, (5) *learned products* and (6) *hybrids*. The last group, 'hybrids', is a combination of strategies from other groups in which students use more than one computational resource. In each strategy group there are different strategy variations. These variations depend on different modes of representation and the techniques that can be used for computational resources. For example, in the strategy *count by*, three variations are presented, according to whether the measurement is made with a full drawing, with written groups or by using fingers.

Based on the above literature presentation of the analysis and categorisation of multiplication tables and strategies, we will present a proposal for the categorisation of strategies in multiplication tables. We propose another classification for the following reasons: We believe that the categorisation of Sherin and Fuson (2005), presented above, contains too many categories, becoming therefore quite difficult for teacher training and application in practice. Strategies used depend each time on the age of the students and on the instruction applied. The two groups of strategies *count all* and *count by* can be combined into one group because they are based on the same counting procedure. We call this group *counting strategies*. We also consider that the pattern-based strategy does not need to be a separate group of its own and can be combined into a group of strategies, called *multiplication strategies*. In multiplication strategies can also be included *derived product strategies*, which Sherin and Fuson present in the category *hybrids*.

We conclude, therefore, with three basic categories of strategies: *counting strategies*, *addition strategies* and *multiplication strategies*, which we present in Table 4.1. A fourth category of strategies is *hybrids*, i.e., strategies that result from the combination of these three main groups of strategies.

4.1.3.1. Counting strategies

This group of strategies includes strategies in which students use counting to find the result of the multiplication. If counting is done one by one, we have the strategy *counting all*. Various sub-classes of this strategy may be noted in the classification of Sherin and Fuson (2005). When counting is done by units greater than one,

TABLE 4.1 Strategies for calculating the multiplication tables

Counting strategies	
Counting all	Here, the students count one by one to find the result of the multiplication. For example, for 3×3: one, two, <u>three</u>, four, five, <u>six</u>, seven, eight, <u>nine</u>.
Rhythmic counting	Students count using a factor of the product as a counting unit. For example, for 4×5: 5, 10, 15, 20.
Addition strategies	
Repeated addition	Students add successively one factor of the product as many times as the other factor is large. For example, for 4×6: $6 + 6 + 6 + 6 = 24$.
Additive doubling	Students double the one factor and add the double to find the result. For example, for 4×6: $12 + 12 = 24$.
Multiplication strategies	
Direct retrieval	Students know the product by heart and retrieve it from the memory. For example, $6 \times 8 = 48$.
Derived product	Students retrieve from the memory learned products, with which they calculate to find the final result. For example, for 6×9, they find the $6 \times 10 = 60$ and subtract 6, $60 - 6 = 54$.
Hybrids	There may be strategies presented that are combinations of the above three categories.

which is one of the factors of the product, we have the strategic category *rhythmic counting*. These strategies are called by Sherin and Fuson *counting by sequence*.

4.1.3.2. Addition strategies

These include strategies in which addition is used in order to find a product. We have two subcategories, *repeated addition* and *additive doubling*. Additive doubling is termed *collapsing groups and adding* by Sherin and Fuson (2005). The strategy *additive doubling* is the way the ancient Egyptians multiplied.

4.1.3.3. Multiplication strategies

In this group of strategies are included strategies in which products of the multiplication tables are used. When students know the products and retrieve them immediately from long-term memory, we say that we have a strategy of *direct retrieval*. When students retrieve from their memory one or more products and calculate the results of the multiplication using them, we say that we have the strategy *derived product*. Strategies of *derived product* are included by Sherin and Fuson (2005) in the group of strategies called *hybrids*.

4.1.3.4. Hybrids

The group of strategies called *hybrids* may be strategies that combine strategies of two or three basic groups of strategies. For example, this might be *direct retrieval + counting all*, or another such combination.

4.1.4. Variations in the use of strategies and factors that affect performance in multiplication tables

4.1.4.1. Differentiation in the use of strategies over time

Research shows that children seem to acquire and use a variety of strategies when solving simple multiplication problems. Siegler (1988) found that children in the second and third year of school used retrieval in 68% of trials, repeated addition in 22% of trials, writing down the problem in 5% of trials and counting object sets in 4% of trials. Lemaire and Siegler (1995) found that the number of strategies used by children in the second school year in France begins with 3.1, rises to 3.7 and then decreases to 2.4. Cooney et al. (1988) identified that children in their third or fourth school year applied retrieval in 55% and 74% of trials, respectively, and used repeated addition, derived product (e.g., they solved 9×6 as $[10 \times 6] - 6$) and other processes (e.g., rules) in the remaining percent. Consequently, at least when capturing multiplication facts, children have multiple paths available leading to the solution (in LeFevre et al., 1996).

Through the study of the development of strategies used in solving simple multiplication tasks, it is shown that procedural strategies are gradually replaced by immediate retrieval from memory. This strategy appears during the early school years and progresses gradually over the following years by the storage in memory of more and more arithmetic facts (Koshmider & Ashcraft, 1991; Siegler, 1988; Siegler & Shrager, 1984). Until higher education, most central arithmetic facts have probably been dealt with so often that the retrieval from memory has become the dominant strategy (Campbell & Xue, 2001).

However, studies pertaining to solving simple arithmetic multiplication problems in adults have shown that experience does not necessarily lead to the exclusive use of the strategy of fact retrieval. LeFevre et al. (1996) find that adults use immediate retrieval in 80% of the trials, but they also use repeated addition, rhythmic counting and derived product.

Lemonidis, Tsakiridou, Panou and Griva (2014) examined 50 prospective teachers in Greece and asked them to mentally solve 12 multiplication table operations using small and easy numbers (6 × 6, 3 × 8, 5 × 9, 5 × 7, 4 × 9 and 7 × 7) and large numbers (8 × 9, 7 × 8, 6 × 7, 6 × 9, 9 × 8 and 9 × 9). The result of this research showed that the most difficult products were 6 × 9 and 7 × 8, which could not be calculated by 22% and 18% of prospective teachers, respectively, and could not be immediately retrieved from memory by 36% and 26% of them. Also, 22% of the prospective teachers could not calculate two out of the 12 given products. It was observed that the dominant strategy used was that of direct retrieval from memory, by 64% to 98% of the students. Other strategies used alternatively, at much lower rates however, were the strategy of the derived product, which was used by 8% of the prospective teachers, and the strategy of reciting the whole multiplication column, which reached 10% for the difficult product 6 × 9. Products on the multiplication tables with small numbers showed a higher success rate and a shorter response time.

4.1.4.2. Diversification in the use of strategies according to the factors of the product

Further, we will examine how the use of strategies is affected by the kind of numbers in the factors of the multiplication tables. In the literature, the following three factors have been identified, which seem to have a significant influence on the answers given by students in the multiplication tables.

1. *Problem size effect*. It is easier for both students and adults to calculate products with small numbers. As the numbers of the products grow, the difficulties and mistakes of students or adults increase (Campbell & Graham, 1985, LeFevre et al., 1996).
2. *Ties effect*. Products that contain doubles (i.e., the two factors are the same, e.g., 7 × 7) are easier for students than the products of corresponding numbers that are dissimilar, e.g., 7 × 6. Response times are longer for products with large numbers (Campbell & Graham, 1985, LeFevre et al., 1996).
3. *Effects of 5-operand problems*. Products having at least one factor of 5, such as 5 × 8, are easy for students and are solved more quickly than other products of comparable size, such as 6 × 7 (Campbell, 1994; Campbell & Graham, 1985; LeFevre et al., 1996).

According to the structural models of multiplication, the characteristics of the numerical relationships between the products, such as the size or relative position of the factors, are accordingly illustrated with their mental organisation (Campbell, 1995; Miller, Perlmutter & Keating, 1984, cited in Penner-Wilger, Leth-Steensen & LeFevre, 2002). For example, products with larger numbers are retrieved more slowly than products with smaller numbers, take longer to resolve and lead to more mistakes, because their representations are less differentiated in memory. In contrast, Siegler's (1988) learning model explains that the problem of the size effect is based on more experience with smaller factors known to appear more frequently in children's textbooks (Ashcraft & Christy, 1995). As a result of this experience, stronger connections between small factors and their correct solutions are formed (Steel & Funnel, 2001).

LeFevre et al. (1996) claim the size of factors has an effect when solving problems, with different options for procedures followed for small and large factors, since many non-retrieved processes take longer to perform than those solved by using retrievals. For example, solving the problem 6×7 using the repeated addition of 6 is a slow process and prone to errors. Similarly, the decomposition of 9×8 to $(10 \times 8) - 8$ is slower than retrieving 72 from memory. Consequently, the effect is likely due to the use of a larger variety of processes for greater factors than small.

De Brauwer, Verguts and Fias (2006) examined 9, 10 and 11-year-olds and adults and showed that the gradual decrease in problem size effect ends already in the sixth grade. The same authors found that the tie and five effects were robustly observed from fourth grade onward and were stable from sixth grade on.

4.1.5. The teaching of multiplication tables

According to what we have seen so far, we can say that the development of strategies for multiplication tables, and generally learning the multiplication tables, is not driven by a universal process of conceptual development. Children's behaviour and performance, as well as that of adults, are diverse and influenced by various factors. A basic factor is the way the teaching of the multiplication tables is done. Other factors are: students' prior knowledge and available computational resources, as we saw above, and idiosyncratic factors, such as the operation and the capacity of short-term and long-term memory, socio-cultural factors, etc.

The traditional way of teaching multiplication tables, according to which students are asked to memorise and repeat (in order) the columns of the multiplication tables is not effective. Students who can recite by heart the columns of the products of multiplication tables do not necessarily know or are not necessarily able to find independent and isolated multiplication facts (Ter Heege, 1985, p. 378). The teaching that emphasises memorising individual arithmetic facts of the multiplication tables is also not appropriate. That is, students who receive instruction with continuous practise in memorising the facts of the multiplication tables do not necessarily learn and memorise the multiplication tables.

We will see in what general contexts the learning of multiplication facts evolve and what strategies students use to calculate products that they do not know by heart.

According to Hans Ter Heege (1985), the learning of multiplication facts is developed in three stages:

In the *first stage*, *conceptual*, an attempt to develop a broad conceptual base for the multiplication operation is made, through experiences from a variety of multiplicative conditions. These conditions are presented in various forms, such as material, written, verbal and symbolic, causing a variety of informal counting and grouping strategies.

In the **second stage**, *reconstructive*, students are encouraged to discover and discuss strategies used to calculate unknown products, based on already known facts. Such strategies are, for example, strategies of counting or repeated addition of calculations based on learned products. Since most children cannot directly memorise the multiplication tables through exercise and practise, they go through an intermediate stage in which they use spontaneously some strategies to produce multiplication facts.

Students gradually end up in the ***third stage*** of the *acquisition* of multiplication tables, *reproduction* and *stabilisation*, after many exercises and discussions focused on these strategies. In this stage, memorisation is emphasised, but students can always resort to the reconstructive process if reproduction fails.

These three stages of the development of multiplication facts are not independent and separated from each other.

According to this logic of the three stages of development, enough time needs to be dedicated at the beginning for the involvement of children in everyday empirical situations of multiplication and division so that these operations become meaningful for them. Next, the opportunity needs to be given to children to operate reconstructively, namely to calculate unknown products based on already known facts and thereby to discover and discuss new strategies. Finally, an emphasis has to be given on memorising and practising the multiplication tables.

Several researchers (e.g., Ter Heege, 1985; Askew & Wiliam, 1995), argue that 'knowledge by heart' or 'memorisation' and 'calculation' are two complementary aspects of the development of pupils' progress in mathematics. Students need to gradually learn more and more arithmetic facts by heart, but it is also essential to develop the opportunity to use prior knowledge to calculate facts new to them or facts they do not remember. Besides, students are more likely to be able to remember arithmetic facts when they feel the confidence and certainty that comes from knowing that they can always calculate a fact that they have inadvertently forgotten.

Ways in which teachers help their students learn the multiplication tables are particularly important, but all subsequent attempts to revise and enhance, or even regenerate this early learning, both in primary and secondary education, also have crucial importance. These efforts need to include the learning of useful calculation strategies that have more general applications, as well as to strengthen links between the facts, rather than to give support to the idea that multiplication tables are a set of isolated facts. Simple admonitions to learn the multiplication tables, combined with frequent examination, may be counterproductive strategies, since they often reinforce failure and do not bring improvements in attitudes and motivation of students. Therefore, more imaginative and creative approaches are demanded for the teaching of the multiplication tables (French, 2005).

Students may use a variety of strategies to calculate arithmetic facts of the multiplication tables. This presentation and discussion of these different strategies in the classroom are very useful. For example, for the product 7 × 8, there may be applied different strategies from the students' side. The most basic strategies are to recite the multiplication table of 8 or 7 to reach the product 7 × 8 or 8 × 7. This product can be calculated from the learned products 7 × 7 or 8 × 8 adding 7 units or subtracting 8 units respectively. The learned product 6 × 8, with the addition of 8 units, can also be used. Multiplying by 8 can be seen as doubling three consecutive times (2 × 7 = 14, 2 × 14 = 28 and 2 × 28 = 56). It can also be used with the addition of already known results (5 × 8 = 40 and 2 × 8 = 16, so 40 + 16 = 56), as well as the method of doubling and halving of the two factors respectively (7 × 8 = 14 × 4 = 28 × 2 = 56).

Several researchers (such as Ter Heege, 1985; French, 2005) have recorded and reported the calculation strategies that students use to calculate a product based on other products they already know. These strategies are:

1. *The commutative law.* Knowing this law, although not necessarily its name (e.g., 8 × 6 = 6 × 8), is very useful in the calculation of the facts of multiplication tables. Thanks to this property of the multiplication tables the number of facts is limited to half.
2. *The distributive law.* For the application of this property, it needs to be understood that any number can be divided in various ways (e.g., 8 × 9 = 5 × 9 + 3 × 9 = 45 + 27 = 72 or 8 × 9 = 4 × 9 + 4 × 9 = 36 + 36 = 72). This property leads to a simplification of multiplication and is a key to understanding algebra.
3. *The inverse nature of multiplication and division.* Understanding the inverse nature of multiplication and division reduces the number of arithmetic facts to be learned. For example, when one of the following three facts is known: 3 × 6 = 18, 18 ÷ 3 = 6 and 18 ÷ 6 = 3, the rest of the facts can be calculated immediately. According to this relationship, divisions are calculated based on multiplication.
4. *Doubling and halving.* Since the multiplication table of 2 is easy for most students, it is not difficult to learn how to mentally double single-digit or double-digit numbers. Multiplications by 4 and 8 can be done by doubling, since these are the second and third power of 2. For example: 2 × 7 = 14 → 4 × 7 = 2 × 14 = 28 → 8 × 7 = 2 × 28 = 56. Also, multiplication by 6 can be accomplished by doubling the product of number 3. For example: 3 × 7 = 21 → 6 × 7 = 2 × 21 = 42. Halving is implemented by students less often. It is used for products of 10 to find products of 5, e.g., to find 5 × 7, 10 × 7 = 70 is calculated and divided in half 70 ÷ 2 = 35.
5. *Multiplying by 10.* This is very easy for students, e.g., 7 × 10 = 70, by adding a zero. Products of 10 may be a benchmark to calculate other products, such as 9 or 5. For the calculation of the products of 9, we subtract once,

e.g., $9 \times 8 \to 10 \times 8 = 80 - 8 = 72$. For the calculation of the products of 5 we divide products of 10 in half.
6. *Multiplying by 5.* If students practise counting by 5, i.e., 5, 10, 15, 20, 25, . . . , products of 5 are learned relatively easily. Products of 5 can be a benchmark for the calculation of other products, such as 6 or 4.
7. *Increase or decrease once.* Students use familiar facts to calculate others that are one position above or below the column in the multiplication tables. As mentioned above, based on facts for multiples of 10 facts of 9 can be calculated, and based on facts of 5 facts of 6 and 4 can be calculated. They may also use other familiar facts, such as $6 \times 6 = 36$ and $6 \times 8 = 48$, which can be memorised easily because of their phonological regularity. For example, $6 \times 7 \to 6 \times 6 = 36 + 6 = 42$.

4.1.5.1. Instructive observations and directions for teaching multiplication tables

According to what has been presented above, we would like to make some observations and give simple directions to teachers for teaching and learning the multiplication tables.

A first general observation we can make is that learning the multiplication tables is a lengthy process. Until students attain the memorisation of many facts they will go through different learning processes: mainly they will go through the constructive stage, in which facts are calculated. For this reason, the teaching of multiplication tables extends over two school years (the second and third grade). Teachers and parents should not expect quick or immediate learning outcomes and compress the time of teaching and learning the multiplication tables.

As mentioned above, according to the first stage of Ter Heege (1985), long before students are introduced to the operations of single-digit multiplications, there should be enough time devoted to teaching examples in everyday life that are familiar to them, so that they understand the meaning of the multiplicative operation and 'how many times'. Chronologically, this is placed in kindergarten and first grade. During this time period counting activities with units bigger than 1 may be introduced. That is, providing students with activities in order to count by 2, by 5 or by 10. As we have seen, this knowledge is one of the first computational resources for calculating the facts of the multiplication tables.

As it is done in today's teaching of multiplication tables, the tables of 2, 5 and 10 are introduced first. These facts are easy for students and very useful because they are a benchmark according to which other facts are calculated.

We saw that students will calculate the facts of the multiplication tables that they have not memorised for a long time. This process reaches to adulthood for those who, having forgotten a fact of the multiplication tables, calculate it using other known facts. So, at first, students, in order to calculate a product, use counting or repeated addition. Then, after memorising some easy facts, these are then used to calculate others. This stage and the process of calculation and memorisation is the most crucial for learning the multiplication tables. With an on-going process of calculation and memorisation of the facts, students memorise more and more facts and also acquire more advanced calculation strategies.

Teachers, therefore, should devote enough time to this stage and treat it with skill. They should know the strategies that students use and observe the way they handle them.

As we saw earlier, the phonological articulation of the facts affects and aids memorisation. This fact can be used in teaching in different ways. The facts of the multiplication tables can be read aloud briefly, so that they can be more easily stored in memory. For example, instead of saying 'three times seven equals twenty-one', we can say 'three seven twenty-one'. This ties facts n × n (2×2, 3×3, 4×4, ...) together with other facts in which a number is repeated, such as $3 \times 5 = 15$, $5 \times 7 = 35$, $5 \times 9 = 45$, $6 \times 8 = 48$, are spoken aloud for easier memorisation.

The instructional method by which the teacher handles the class is similar to that applied to other mental calculations. It gives students opportunities to speak aloud or write the way products are calculated. This action of students is meta-cognitive and benefits learning a great deal. Also, students have the opportunity to interact with new strategies and generally learn the way their classmates think. Collaborative discussion takes place in the classroom, and various ways of calculating are presented so that all the students can see them. There are a variety; there is pluralism in strategies and in the ways students calculate. We do not push any particular way, allowing students to use individual methods suiting their idiosyncrasy.

4.2. Division

It is known that division is an operation that students often find more difficult than the other operations. The operations of division and multiplication are the so-called multiplicative relations. Multiplication and division are conceptually complex to the extent of semantic structures, as we will see below, but also in conceptual understanding (Steffe, 1988). As discussed in Section 1.3.1, multiplicative relations are characterised by the fact that they are based not only on simple units that are counted, as it happens in the case of addition and subtraction, but also in groups of equal multitude which should be considered as a unit. For example, in the product 3×6 we have three groups of six elements, in which six elements can be considered together as one unit. The consideration of multiple units as one unit is a higher-level abstraction than the thought of a single unit (Clark & Kamii, 1996, pp. 42–43).

According to modern programmes, in elementary school, division is introduced from the first grade through the use of informal strategies in sharing activities with objects. Then, in the second and third grade, students practise division problems and situations and perform mental divisions. The standard written algorithm of division is introduced either in the last grade of elementary school or in the first grades of middle school. In the Netherlands, the standard algorithm of division is not taught in primary school (see 2.2.4.4.). In old programmes, the traditional logic was followed most intensely, whereby great emphasis was placed on standard written algorithms, which is why division was imported earlier, before students were able to perform mental calculations.

Division is the last of the four arithmetic operations that the child learns. Divisions are the most difficult operations, with respect to being learnt or automated, probably because they are given relatively little practise in relation to the other operations (Siegler & Shipley, 1995). The other operations are commonly used as a basis for the learning of division (Parmar, 2003). Thus, the child's performance in the division may be affected by how well he has learned the previous operations.

4.2.1. Semantic structure of multiplication and division problems

Vergnaud (1983, 1988) identified the complexity that exists in relation between the contents of word problems and the way the operations of multiplication and division are used for their modelling. He identified a large context, which calls 'the conceptual field of multiplicative structures', and proposes four major categories of multiplication and division problems, which are: (a) *isomorphism of measures*, (b) *multiplicative factor or product of measures*, (c) *Cartesian product* and (d) *multiple proportion*.

Greer (1992) also presents four categories, mainly applied to problems that involve multiplication of whole numbers:

- equal groups (e.g., 3 tables, 5 people at each);
- multiplicative comparison (e.g., there are three times as many boys as girls);
- rectangular rows (e.g., 5 rows of 3 children);
- Cartesian product (e.g., the number of possible combinations of a skirt and a blouse).

Each multiplicative condition leads to various division problems. For example, in equal groups we have the categories of partitive division (the total number of people is 15, there are 3 tables, how many are at each table?) and quotitive division (the total number of people is 15, there are 5 people at each table, how many tables?).

According to studies, mathematically equivalent problems with different semantic structures cause different solution strategies and vary greatly in their difficulty (Fischbein et al., 1985; Vergnaud, 1983). Mulligan and Mitchelmore (1997, p. 310) note that classifications based on semantic structure are arbitrary and that categories can be expanded, undone or improved in accordance with the requirements of research.

4.2.2. Partitive and quotitive divisions

Vergnaud (1981, 1983) and Fischbein et al. (1985) propose two types of division: *partitive division* and *quotitive division*.

Partitive division refers to the concept of sharing. The division in the following problem is partitive division: 'The grandmother wants to divide 45 Euros equally among her three grandchildren. How many Euros will she give to each one?' In partitive divisions, we know the value corresponding to the many units (45 Euros), the number of units (three grandchildren) and we want to find the value that corresponds to one unit (how many Euros will she give to each one?). So, we divide (share) the price of the many units by the number of these units (45 ÷ 3). In partitive divisions the quotient (i.e., the objective) is always a magnitude similar than the dividend. For the example we divide 45 Euros by 3 grandchildren, finding 15 Euros per grandchild.

In the following problem the division is *quotitive division*: 'The grandmother has 45 Euros and gives to each grandchild 15 Euros. To how many grandchildren has the grandmother given money?' In quotitive divisions, we know the value of one unit (each grandchild gets 15 Euros) and the value of many units similar to this (45 Euros) and we want to find the number of these units (how many grandchildren got money). In quotitive divisions, we divide two values of similar magnitude (45 Euros and 15 Euros) and we find a quotient expressed in another magnitude (3 grandchildren).

Partitive division is taught before quotitive division, because the concept of sharing is considered to be a concept that students encounter in their everyday lives. Brown (1992), however, finds that students in second grade perform better in quotitive problems and solve sharing problems using grouping strategies rather than sharing strategies. Other research has shown that long-term, the aspect of sharing in the division is more restricted than the quotitive aspect (Correa, Nunes & Bryant, 1998; Haylock & Cockburn, 1997).

4.2.3. Initial knowledge of students in division

Sharing is an informal daily activity that plays an important role in the initial understanding of division. Sharing is division, with the meaning that to share an amount successfully, you have to divide a dividend into equal amounts. At school, division is usually introduced as a form of sharing, because this is an action children are able to perform.

Studies show that most four-year-olds and five-year-olds know how to share amounts in a distribution, in this way: they distribute sequentially one to A, one to B, one to A, one to B, etc. until the dividend ends. It would also seem that around five years of age, children can already understand several things according to the basis of this process (Miller, 1984; Frydman & Bryant, 1988). It has also been found that small children can model division problems using materials before they are typically taught division (Carpenter et al., 1993; Correa, 1994), and that the initial strategies of children show the action described in this problem (e.g., Marton & Neuman, 1996). This facility of students in sharing means that this activity fits with the *action schemas* (Piaget, 1972/1947) according to which the understanding of division can be developed.

Understanding division requires more than knowledge of how to share a collection in equal parts, it requires awareness of the relationship between a divisor and the quotient (Bryant, 1997). That the amount received by each varies according to the number of recipients is not realised by many young children when they share. Bryant argues that a small child may be able to share using one-to-one correspondence, but it is unlikely that he has an understanding of this relationship.

4.2.4. Categorisation of strategies in simple divisions

In this section, we will examine the categorisation of strategies that students use in simple divisions. When we say simple divisions, we mean single-digit divisions, i.e., two digit by single digit and generally those that can be solved with the one step of performing a simple multiplication with known products (e.g., 12 ÷ 4, 32 ÷ 8, 85 ÷ 9, etc.).

Over the past few decades research has been conducted to study the strategies used by children when performing the operation of division or multiplication and division together (Brown, 1992; Downton, 2008; Kouba, 1989; Mulligan & Mitchelmore, 1997; Oliver, Murray & Human, 1991; Robinson, Arbuthnott, Rose, McCarron, Globa & Phonexay, 2005; Steffe, 1988). These studies conclude that the strategies used by children generally begin with direct modelling and unitary counting, proceed to skip counting, double counting, repeated addition or subtraction, then use known multiplication or division facts and, finally, commutativity and derived facts. At first, direct modelling is used to solve problems, and then multiplicative thought with partial modelling is developed, eventually ending with an abstract way of operation when faced with problems. Before this point they are unable to develop or integrate complex structures with counting strategies.

Mulligan and Mitchelmore (1997) find that children gradually develop a sequence with increasingly more efficient intuitive models. New models come from earlier models; children do not switch from one model to the next, but develop a growing range of models, on which they base the solution of a problem. In the following, we present the intuitive models and the corresponding strategies that children develop for division, according to Mulligan and Mitchelmore (1997, p. 316).

1. *Direct counting*: One-to-many correspondence. Unitary counting. Sharing. Trial-and-error grouping.
2. *Repeated subtraction*: Rhythmic counting backward. Skip counting backward. Repeated subtracting. Additive halving.
3. *Repeated addition*: Rhythmic counting forward. Skip counting forward. Repeated adding. Additive doubling.
4. *Multiplicative operation*: Known multiplicative fact. Derived multiplicative fact.

Direct counting. Objects are simply counted without any apparent reference to the multiplicative structure. For example, I share 12 biscuits fairly among 3 children: I give each child one in order, then one again, until all cookies are done. After sharing I count how many cookies each child received.

Rhythmic counting. Counting follows the structure of the problem. Simultaneously with counting, a second counting of the number of the groups takes place. For example, I have 12 cookies and give 4 to each child, how many children will get cookies? Of the 12 I subtract 4 and I have 8 left, of the 8 I subtract 4 and I have 4 left, how many times did I get 4?

Skip counting. Counting is done with multiples. For example, how many fives are there in 20? 5, 10, 15, 20, and simultaneously I count the steps, 1, 2, 3, 4.

Repeated addition or subtraction. Counting is replaced by calculations such as $3 + 3 = 6$, $6 + 3 = 9$ or $9 - 3 = 6$, $6 - 3 = 3$.

Additive doubling. For example, in operation $16 \div 4$, '4 and 4 equals 8, 8 and 8 is 16'.

Additive halving. For example, in operation $16 \div 4$, 'if we cut 8 into two halves, we have 4 and 4'.

Multiplicative operation. The calculation is done by retrieving from memory a known multiplication fact or by production of a known fact: $12 \div 3 = 4$ because $3 \times 4 = 12$.

4.2.5. Study of simple operations of division

Multiplication and division are two complementary mathematical operations. If someone computes the division $54 \div 6 = ?$ he would most possibly think of the corresponding multiplication $6 \times 9 = 54$ to find 9. It seems therefore, that to calculate division many times, we retrieve from memory and use the inverse operation of multiplication. This leads to the intuitive assumption that multiplication and division facts (e.g., $6 \times 9 = 54$ and $54 \div 6 = 9$) are represented together in memory.

Many studies have investigated how these facts are related to one another in memory (e.g., De Brauwer & Fias, 2011; Campbell, 1997, 1999; Campbell & Robert, 2008; LeFevre & Morris, 1999).

A theory which interprets representation of arithmetic fact in memory is the *Identical Elements (IE) model* (Rickard, 2005; Rickard & Bourne, 1996; Rickard, Healy & Bourne, 1994). According to this model, for arithmetic facts composed of identical elements (i.e., operands and answer) regardless of operand order, there is a single memory node. For example, each multiplication node (e.g. [6, 9, × → 54]) is accessed by either operand order (6×9 or 9×6). Problems that present different operands access different nodes. For example, the inverse division problems (e.g., $54 \div 9$ and $54 \div 6$) are represented by different nodes. Regarding the transfer of learning, the prediction which the IE model makes is that the transfer will be restricted to problems with identical elements.

Campbell (1999) found a transfer from multiplication to division that was interpreted as evidence of a *mediation strategy*. This means that large division problems (e.g., $63 \div 7 = ?$) are not solved by direct retrieval, but by retrieving the corresponding multiplication fact (e.g. 9×7 or 7×9) from memory.

To account for mediated transfer (division by mediation), Rickard (2005) introduced a *revised IE model (IE-r)*. This IE-r model specified the mechanism by which multiplication mediates division. At intermediate skill levels, in the non-asymptotic case of the model, there may be a reverse association between operands and product (e.g., 6, 9, × → ← 54) for multiplication problems. This means that products can activate their operands (factors) and vice versa.

The results of De Brauwer and Fias's (2011) research results showed that memory and learning processes do not seem to differ fundamentally between addition–subtraction and multiplication–division. They observed retrieval savings between inverse multiplication and division problems. Even for small problems (solved by direct retrieval) practising a division problem facilitated the corresponding multiplication problem and vice versa. According to the authors these findings indicate that shared memory representations underlie multiplication and division retrieval.

Research results with adult participants were consistent with the revised identical elements model (De Brauwer & Fias, 2011; Campbell, Fuchs-Lacelle & Phenix, 2006; Rickard, 2005; Rickard & Bourne, 1996; Rickard, Healy & Bourne, 1994).

De Brauwer and Fias (2009) measured the performance of 8-year-old children longitudinally twice a year to determine the developmental trajectories of simple multiplication and division. To investigate the relationship and the developmental parallels in performance of multiplication and division, they observed all effects (problem size, 5 and tie effect and tie × size interaction) in multiplication and division.

Their results indicate strong developmental parallels between both operations. These results are expected for strongly interconnected memory networks for multiplication and division facts, at least in young children.

In Robinson et al.'s (2005) research the behaviour of Canadian students from fourth to seventh grade in simple division problems was examined. The main research questions were: What are the strategies that students use in simple division problems and how do these strategies change and develop over time? Another question was: Is there is a problem size effect in divisions?

The three most common strategies which appeared in all classes were the direct retrieval of the division of memory (e.g., 20 ÷ 4, 'is 5, I know it'), multiplication (e.g., 20 ÷ 4, ? × 4 = 20) and addition (e.g., 20 ÷ 4, 4 + 4 + 4 + 4 + 4).

The use of these strategies vary depending on the grade. Fourth-grade students used addition more, while the use of this strategy decreased as the grade level increased. Multiplication in fourth grade was used less, but in older grades it was used more. For the strategy of retrieving from memory, a strange phenomenon occurred. The percentage of students using this strategy was low (about 16%) and remained almost stable as grades increased. It was observed that, while in the other three operations (addition, subtraction and multiplication) as the grades increased the percentage of students using the strategy of direct retrieval also increased, for the operation of division the percentage of using this strategy was low and remained almost stable, i.e., did not increase in higher grades.

In the research of Robinson et al. (2005), it was found that there is a problem size effect on students' behaviour in simple divisions. Students are generally quicker and more correct in their divisions when they are operating on small numbers than with large numbers.

Walker, Bajic, Mickes, Kwak and Rickard (2014) investigated delayed transfer effects among children, 6 to 11 years old, trained for six sessions on either a set of mixed addition and subtraction problems or a set of mixed multiplication and division problems. In this research the following two major empirical questions were posed: are there performance improvements in the transfer from trained problems to untrained problems with the same operation? Compared with untrained problems, is there a greater transfer of learning to operand-inverted problems? The results show that across all ages and for both pairs of arithmetic operations, the substantial learning that was observed for trained problems did not transfer to either operation-inverted or untrained problems. The authors concluded that the specificity of learning observed in this study is nearly identical to that observed among adults. A similarly high degree of learning specificity among children was observed by Walker, Mickes, Bajic, Nailon and Rickard (2013).

4.3. Multi-digit multiplications and divisions

4.3.1. Classification of strategies in multi-digit multiplications and divisions

In Section 4.1.3 and Section 4.1.4, the categorisation of strategies for calculating in multiplication tables was presented, and in Section 4.2.4 the categorisation of strategies for simple divisions was shown. As we saw in these sections, there are several studies that examine students' strategies for simple multiplications and divisions. However, for multi-digit or double-digit multiplications and divisions, there are not many studies in the bibliography concerning strategies that students use (Baek, 1998; Heirdsfield, Cooper, Mulligan & Irons, 1999; Murray, Olivier & Human, 1994; Trachilou, Christou & Lemonidis, 2008).

Baek (1998, pp. 151–160) carried out a research on students of the third to the fifth grades of elementary school, in order to investigate which multiplication algorithms they would invent (invented algorithms). The students examined had never been taught any rules or typical algorithms. Students solved problems posed to them containing multi-digit multiplications individually or in groups. Pupils discussed and compared the algorithms they used with each other. In this study, the algorithms invented by the students for multi-digit multiplication problems were classified into four categories: *direct modelling*, *complete number strategies*, *partitioning number strategies* and *compensating strategies*.

Heirdsfield et al. (1999) investigated and classified strategies in multi-digit multiplications and divisions used by students of fourth to sixth grade in Australia. At multi-digit multiplication strategies, those researchers do not mention the modelling strategy, as Baek (1998) above, but they mention as a first strategy the *counting* strategy. For this counting strategy, as they call it, they include any form of counting strategy, skip counting forwards and backwards, repeated addition and subtraction, and halving and doubling strategies. These authors refer to the compensation strategy as a *holistic strategy*.

Based on Baek's (1998) strategies and classification, the classification and the terms of the strategies of Heirdsfield et al. (1999), terms used internationally for strategies and, finally, our research experience in applying a classification of strategies, we propose the classifications of strategies in multi-digit multiplications and divisions in Table 4.2.

TABLE 4.2 Classification of strategies in mental multi-digit multiplication and division

Strategy	Description	Example
1. Direct modelling	Students model the problem and count the total number of objects, the number of groups or the number of the objects in every group.	Multiplication: They count in the model the total number of objects. Division: They count in the model the number of groups (quotitive) or the number of objects in each group (partitive).
2. Counting	Every form of count strategy, skip counting forwards or backwards, repeated addition or subtraction, doubling and halving strategies.	Multiplication: 5×15: 15, 30, 45, 60, 75 or 5×15: $2 \times 15 = 30$, $30 + 30 = 60$, $60 + 15 = 75$. Division: $75 \div 5$: 15, 30, 45, 60, 75 or $180 \div 4$: $180 \div 2 = 90$, $90 \div 2 = 45$.
3. Direct retrieval	They use a known multiplication or division fact or a derived fact.	Multiplication: $8 \times 11 = 88$, $5 \times 12 = 60$. Division: $120 \div 6 = 20$, because $6 \times 20 = 120$.
4. Partitioning number	They partition one or both of the operation's terms at minor numbers, in order to be able to multiply or divide them easier.	
4.1. Partitioning a number based on place value	One number is partitioned based on the place value of the arithmetic system.	Multiplication: $7 \times 15 = (7 \times 5) + (7 \times 10) = 35 + 70 = 105$.

Partitioning RL	The number is partitioned and they act from right to left.	Division: $84 \div 4$: $4 \div 4 = 1$, $80 \div 4 = 20$, $20 + 1 = 21$.
Partitioning LR	The number is partitioned and they act from left to right.	Multiplication: $7 \times 15 = (7 \times 10) + (7 \times 5) = 70 + 35 = 105$. Division: $84 \div 4$: $80 \div 4 = 20$, $4 \div 4 = 1$, $20 + 1 = 21$.
4.2. Partitioning both of the numbers based on place value	Multiplier and multiplicand are partitioned at numbers based on the place value.	$14 \times 26 = (10 + 4) \times (20 + 6) = (10 \times 20) + (10 \times 6) + (4 \times 20) + (4 \times 6)$.
4.3. Partitioning not based on place value	Multiplier or multiplicand are partitioned, not based on the place value of the arithmetic system.	$15 \times 136 = (5 \times 3) \times 136 = 5 \times (3 \times 136)$ or 7×15: $(5 + 1 + 1) \times 15 = 75 + 15 + 15$.
5. Holistic or compensating	Numbers are treated as wholes.	Multiplication: $8 \times 99 \rightarrow 8 \times 100 - 8$ $50 \times 46 = 100 \times 23$. Division: $940 \div 5$: $940 \div 10 = 94$, $94 \times 2 = 188$ $105 \div 15$: $4 \times 15 = 60$, $3 \times 15 = 45$, $60 + 45 = 105$, therefore $3 + 4 = 7$.

1. **Direct modelling:** This category includes strategies in which students, in order to solve a multiplication or division problem, need to accomplish problem modelling. Even with large numbers, some children need to model the whole situation of the problem and to count all objects. Children model the numbers of the problem using counting materials, material groups based on ten, counting signs or other schemes with which they count the numbers of the problem. There are two types of direct modelling of multiplication: direct modelling with units and direct modelling with tens. In the first case, the student models and counts with units to find the total, while in the second case he counts with tens or larger numbers. For example, in the figure below we present a solution with modelling of the multiplication 25 × 21.

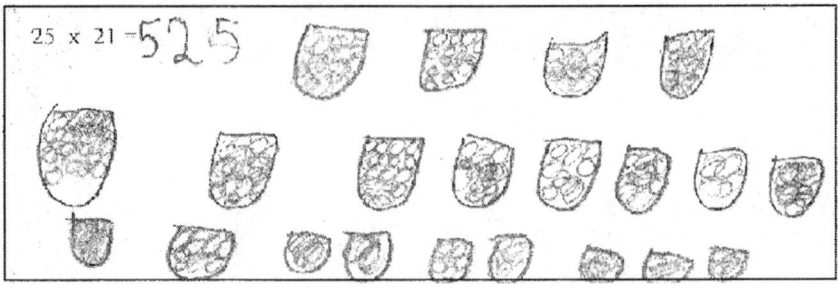

Modelling of the multiplication 25 x 21

In problems of division, modelling is affected by the semantic structure of the problem. For example, for partitive divisions (i.e., I fairly share 450 Euros among 3 children) students create a model in which they share the Euros in three parts and find the amount of Euros each part will contain. For a quotitive division (i.e., I have 520 Euros in coins of 5 Euros. How many coins do I have?), they create a model and try to find how many groups of 5 the number 520 contains.

2. **Counting strategies:** These strategies are called *counting* because students count forwards or backwards or they increase by doubling and they decrease by halving. In these strategies, children add the multiplicand or they subtract the divisor as a whole number, without partitioning the terms of the operation in any specific way. In order to add the multiplicand, children use different strategies, such as repeated addition or doubling. For example, in the multiplication 7 × 15 one child adds 15 seven times, another child doubles 15 and finds 30, adds 30 to 30 and finds 60, then adds another 30 and finds 90, then adds 15 more to 90 and finds the result, 105.

3. **Direct retrieval:** This strategy refers to multiplications or divisions that can be solved with the assistance of direct retrieval by the memory of an arithmetic fact or with a slight modification to an arithmetic fact and the product of an operation with that arithmetic fact. For example, 660 ÷ 11 = 60, because 6 × 11 = 66 and 66 × 10 = 660.

4. **Partitioning number strategies:** In these strategies, children partition one or both of an operation's terms at smaller numbers, so as to be able to multiply or divide them more easily. Children partition one term of the multiplication or the division in two different ways: many children partition numbers based on the place value of the digits: at units, tens, etc.; a few children partition the numbers too, in a way not based on the place value of the digits (non-decade numbers).

   4.1. **Partitioning a number based on place value:** Among these strategies, we can break out two groups: partitioning from right to left (RL) and partitioning from left to right (LR). Here, students use their knowledge of the decimal system in order to partition one of the two terms of a multiplication or a division. In this way, children find the products or the quotients much more easily and use this strategy for many other multi-digit multiplication or division problems. After the partition, in some cases, children operate from right to left. This direction, in the multiplication case, is compatible with the direction of the written algorithm. In other cases, after the partition, children operate from left to right, which, in the division case, resembles the direction of the written algorithm.

   For example, a student uses the strategy of number partitioning in tens to solve the multiplication 25 × 21: He writes the number 25 vertically 21 times and using horizontal lines he splits them into 10. He knows that 10 times 25 gives 250, which he writes in every group of 10. He adds two times of 250 and then adds to them one more 25.

   4.2. **Partitioning both of the numbers based on place value:** This strategy is used only in multiplications: it resembles the standard algorithm. On this occasion, children partition both the multiplier and the multiplicand in numbers based on place value, do each multiplication and then add partial products. For example, a child, in order to calculate 25 × 21, created four partial products: He said that 25 times 21 is 20 × 20, 20 × 1, 5 × 20 and 5 × 1, because 20 times 21 is 20 × 20 plus 20 × 1 and 5 times 21 is 5 × 20 plus 5 × 1. A usual mistake in this strategy is that children, in the case of the multiplication of two-digit numbers, do not form all four partial products; instead, they count only two of them.

   4.3. **Partitioning not based on place value:** This strategy is used only in multiplication. Children who use this strategy partition the multiplier or the multiplicand, so as to make the multiplication easier, or use known products. For example, in the multiplication of 15 × 136 a child considers 15 as 3 × 5. He first counts the product 3 × 136, multiplying the result by 5. The algorithm is equal to the relation: 15 × 136 = (5 × 3) × 136 = 5 × (3 × 136).

   Some children use partitioning not based on place value strategy with the distributive property. For example, in the problem: 'Each one of the 7 dwarves gives Snow White 15 apples. How many apples will Snow White receive?' a child partitions the multiplier 7 into 5 + 1 + 1, and counts: 5 × 15 makes 75, as a result, 6 × 15 makes 90 and 7 × 15 makes 105. Consequently, the answer is 105.

5. **Holistic or compensating strategies:** In these strategies children handle the numbers of the operation in a holistic way. They adjust one or both of the multiplication or division terms, so that the numbers will be doubled or halved, in order to make the calculation easier or to use some multiplication products or quotients that they already know. This strategy is used in multiplications and divisions where numbers have special features. Strategies that adjust both of the numbers are usually used in problems that involve 5. For example, a child, in order to solve the multiplication 50×46, splits 46 in half, to find 23, then multiplies by 100 to find 2,300.

Usually, when children adjust one of the two numbers in a multiplication, they adjust it upwards or downwards to the nearest ten and correct the result afterwards. For example, a child calculates the multiplication 15×48 as follows: $15 \times 50 = 750$, $2 \times 15 = 30$, $750 - 30 = 720$.

4.3.2. Research results of the School of Nature and Life Mathematics in multi-digit multiplications and divisions

4.3.2.1. Research on informal strategies in multiplication before teaching the standard algorithm

Trachilou et al. (2008) carried out a study to investigate the informal strategies that Cypriot students develop in order to solve two-digit multiplication problems before being taught the standard multiplication algorithm. The sample consisted of 75 students, in the third grade of elementary school. Students had been taught the multiplication table, but they had not been taught systematically either the partitioning numbers strategy or the standard multiplication algorithm.

To carry out the survey, an essay was compiled, which included two verbal problems and a pure multiplication. The two verbal problems were the following:

1. Everybody knows the fairy tale with Snow White and the 7 dwarves. So . . . One evening, the dwarves went to collect some apples, in order to bring them to Snow White so that she could make an apple pie. When they came back home, each dwarf had 15 apples in his basket. Snow White was very glad and used all the apples for her apple pie. How many apples did Snow White use in total?
2. Stavros knows about the importance of saving money. For that reason, he saves £12 per day. He wants to calculate how much money he will be able to gather till the end of the month. He considers that a month has 30 days. Can you help him with his calculations?

To solve the first problem, students had to multiply a digit and a two-digit number (7×15), while in the second problem they had to multiply two two-digit numbers, one of which was multiple of 10 (12×30). The pure multiplication, which was the hardest task of the essay, involved the multiplication of two two-digit numbers (25×21).

Researchers urged students to write in an essay the way they thought in order to reach the solution. In addition to the essay, 15 children were personally interviewed at the end of their work, to explain the way they worked to answer the essay. The results of this survey are presented below.

In order to calculate the multiplications given in the three questions, students used the following strategies: the direct modelling, the repeated addition strategy, the partitioning number strategy and standard algorithms.

Students' success in multiplication was 69.5% for the multiplication 7 × 15, 64% for 12 × 30 and 52% for 25 × 21.

According to Table 4.3, many students proceeded to model the first problem. Specifically, they modelled the first problem at a percentage of 34.5%, 10.5% modelled the second one and only 4% modelled the third. We observe that the number of students using modelling in multiplication is reduced noticeably from the first to the third task of the essay, because the numbers are increased. Moreover, the third multiplication was not given in the context of a problem, as were the other two, but as a simple operation, a fact that may have reduced the motivation to model it.

Regarding the repeated addition strategy, the rate of students using this particular strategy was increased in the first and the third task. Students approach the addition strategy as a strategy through which they can reduce mistakes, as they can handle addition as an operation better.

Concerning the partitioning strategy, we find that, on average, 18% of students used this strategy. According to Table 4.2, number partitioning can take place in three ways: (a) *partitioning a number based on place value*, (b) *partitioning both the numbers based on place value* and (c) *partitioning not based on place value*. In this sample of students, none used *partitioning not based on place value*. That could have been expected, as it was too early for these students to partition numbers and operate different individual combinations.

We notice, however, that many students (18%, avg.) tried to partition numbers into tens and units and to perform the partial multiplications, sometimes successfully, while others did not. This ability to partition numbers and create partial products appears as an informal ability, as students had never been taught this.

In cases where students were partitioning both of the numbers, there were many mistakes in this strategy. Students were performing only two partial multiplications, simply multiplying the first digits from both numbers, and did not perform four partial multiplications, multiplying each digit by the other two digits.

Concerning the standard strategy of algorithms a large proportion of students placed numbers vertically to accomplish the operation. This fact can

TABLE 4.3 Success rates and strategies used by third-grade elementary students for two-digit multiplications (from Trachilou et al., 2008)

	Success	*Direct modelling*	*Repeated addition strategy*	*Partitioning number strategies*	*Standard algorithm*
Prob. 1 7 × 15	52 (69.5%)	26 (34.5%)	17 (22.5%)	14 (18.5%)	9 (12%)
Prob. 2 12 × 30	48 (64%)	8 (10.5%)	8 (11%)	17 (22.5%)	12 (16%)
E × er. 3 25 × 21	39 (52%)	3 (4%)	24 (32%)	10 (13.5%)	21 (28%)

be explained by the tendency to imitate and act in the same way as standard algorithms of addition and subtraction. In the third task, the percentage of the use of the vertical algorithm rose (28%); this may be due to student inability to respond to the task using a different strategy, since numbers are increased. Many students who tried to do multiplication vertically completed addition instead of multiplication.

It should be mentioned that many students who used some of the strategies above made errors in the operations, such as failing to multiply some of the numbers in the multiplication. We present such mistakes below, through quotations from the interviews where students explain the way they thought.

> Andrew (Problem 2: 12 × 30): 'I multiplied 2 by 0 and found 0, then I multiplied 3 by 1 and found 3, and therefore the answer is 30 pounds'.
>
> Maria (Exercise 3: 25 × 21): 'I multiplied 5 by 1 and found 5 and then I multiplied 20 by 20 and found 400. 400 + 5 makes 405'.
>
> Katerina (Exercise 3: 25 × 21): 'I multiplied 5 by 1, which gives 5, and 2 by 2, which gives 4, therefore the answer is 45'.

4.3.2.2. Research on strategies of elementary school students after the teaching of standard algorithm

We present the results of the distance training programme for an experimental implementation of mental calculations for in-service teachers of elementary school (see Section 3.5.1). Here we present the results of this survey for multi-digit multiplications and divisions.

As mentioned before, in this survey, teachers of the class made a first examination of the students' performance (pre-test) in mental calculations and subsequently, a second examination with the same questions (post-test) after the teaching intervention. We will present data from the following grades: third (N = 111), fourth (N = 138) and fifth and sixth (N = 308).

Results by pre- and post-test in the third grade

In Table 4.4. below, we present the success rates for different mental multiplications and divisions, and the rates of the strategies used by students before and after the experimental intervention. The rates after the teaching intervention are presented in bold colours.

Success in mental multiplication and division in third grade

Reviewing the first column of the success rates in multiplication and division in Table 4.4 before the teaching intervention, we notice the following: single-digit multiplication of large numbers (6 × 8, 7 × 9) presented high success percentages

TABLE 4.4 Rates of success and strategy use among third-grade students before and after experimental teaching

Operation	Success rate N = 111		Repeated addition or subtraction		Derived product		Direct retrieval	
6 × 8	100	100	3	2	6	4	86	93
	90%	90%	2.5%	2%	5.5%	3.5%	77.5%	84%
7 × 9	95	101	3	2	8	5	78	89
	85.5%	91%	2.5%	2%	7%	4.5%	70%	80%
5 × 12	80	93	6	5	50	41	17	39
	72%	84%	5.5%	4.5%	45%	37%	15.5%	35%
8 × 25	53	78	4	19	46	51	8	5
	47.5%	70.5%	3.5%	17%	41.5%	46%	7%	4.5%
30 ÷ 10	74	102	3	2	29	59	40	36
	66.5%	92%	2.5%	2%	26%	53%	36%	32.5%
54 ÷ 6	64	98	2	2	18	51	39	40
	57.5%	88.5%	2%	2%	16%	46%	35%	36%
48 ÷ 4	61	86	4	5	42	52	17	25
	55%	77.5%	3.5%	4.5%	38%	47%	15.5%	22.5%

(90%, 85.5%), the success rate dropped (72%) in multiplication of single-digits by two-digits with an easy number (5) as multiplier (5 × 12) and it dropped even more (47.5%) in multiplication of single-digit by two-digit numbers with a high multiplier (8) (8 × 25). Division is harder than multiplication: even the division 30 ÷ 10 displayed success only at 66.5%. Divisions 54 ÷ 6 and 48 ÷ 4, which can be related to the multiplications 7 × 9 and 5 × 12, displayed lower success rates. After the teaching intervention, we see in Table 4.4 that, apart from the operations of the multiplication table that already had high success rates, success rates in all other operations were significantly increased. Typical was the case of the multiplication 30 ÷ 10, as the rate increased to 92% from 66.5%.

Use of strategies

Regarding the two facts of the multiplication table (6 × 8 and 7 × 9), we notice that in the third grade, students counted products with large numbers, mainly using the direct retrieval strategy. This means that they have stored these facts and simply retrieved them from the long-term memory. In multiplication and division of single-digit by two-digit numbers, students mainly used the *derived product* strategy, i.e., for the division 54 ÷ 6 they retrieved the product 6 × 9 = 54 and, based on it, they could find the result. In Table 4.4 we see large rates even in the *direct retrieval* strategy, where apparently students could not retrieve the division 54 ÷ 6 directly, but they obviously retrieved the corresponding product 6 × 9 = 54. As these strategies were codified by the classroom's teachers, they codified the same strategy in some cases as *derived product* while in other cases as *direct retrieval*, possibly because there was a retrieval of the reverse product.

TABLE 4.5 Rates of success and strategy use among fourth-grade students before and after the experimental teaching

Operation	Success rate N = 138		Standard algorithm		Repeated addition or subtraction		Derived product (for divisions)		Partitioning number strategies		Holistic or compensating strategies	
8 × 25	100	111	27	13	20	21	3	13	42	43	0	44
	72.5%	80.5%	20%	9.5%	14.5%	15%	2%	9.5%	30.5%	31%		32%
50 × 46	74	94	26	15	0	14	2	8	41	46	2	40
	53.5%	68%	19%	11%		10%	1.5%	6%	30%	33.5%	1.5%	29%
8 × 99	65	94	29	6	0	14	0	11	25	38	19	50
	47%	68%	21%	4.5%		10%		8%	18%	26%	14%	36%
12 × 18	56	79	26	5	1	11		3	31	43	4	30
	40.5%	57%	19%	3.5%	0.5%	8%		2%	22.5%	31%	3%	22%
54 ÷ 6	101	120	24	6	1	16	65	75	0	13	6	30
	73%	87%	17.5%	4.5%	0.5%	11.5%	47%	54.5%		9.5%	4.5%	22%
48 ÷ 4	94	106	27	9	0	7	50	54	11	31	4	30
	68%	77%	19.5%	7%		5%	36%	39%	8%	22.5%	3%	22%
450 ÷ 15	73	83	24	6	2	14	35	35	8	24	4	29
	53%	60%	17.5%	4.5%	1.5	10%	25.5%	25.5%	6%	17.5%	3%	21%

Results by pre- and post-test in the fourth grade

Success in mental multiplication and division in fourth grade

A general comment that we can make, looking the rates of Table 4.5, is that before the teaching intervention students presented low success rates for two-digit multiplications and divisions. Multiplication 8 × 25 and divisions 54 ÷ 6 and 48 ÷ 4 had success rates around 70%. Multiplication of two-digit numbers × two-digit numbers (50 × 46, 12 × 18) as well as single-digit numbers × two-digit numbers (8 × 99), but also the division 450 ÷ 15, were done correctly by almost half of the students. We can state that students' ability in mental calculations of two-digit multiplications and divisions is too low.

After the teaching intervention, there was an increase in rates of success that is statistically significant in all questions, except for the operations: 8 × 25 and 450 ÷ 15, in which there was also an increase of the success rate, but only within the limits of statistical significance. It seems that there were amelioration conditions for students. Thus the experimental teaching intervention produced an increase of students' ability to mentally calculate two-digit multiplication and division.

Use of strategies

Concerning the strategies that students used before and after the teaching intervention, we observe a significant alteration in their use. We noted in Table 4.5 that before the teaching intervention, students used the standard algorithm at a rate around 20%, they rarely used repeated addition or subtraction, apart from the case of the operation 8 × 25, where it was used by 14.5% of the students. They used the strategy of the partitioning number in multiplication, rarely using holistic or compensating strategies, apart from the operation 8 × 99, which was only used by 14% of the students. This illustration of the use of strategies by students shows that they may not have been trained at all in mental calculation.

After the teaching intervention, students used different strategies. The use of the standard algorithm decreased a great deal, such that the rate of students using the standard algorithm dropped from an average of 19% to 6.5%. Also, the use of the repeated addition or subtraction strategy, and the doubling or halving strategies appeared at 10% on average, while, before the teaching intervention, they were never used, apart from in the case of 8 × 25. Holistic or compensating strategies were used after intervention at an average of 26.5%, while before the teaching intervention they were rarely used. Through teaching intervention, students decreased their use of the standard algorithm and increased the use of advanced and flexible strategies, as well as holistic ones, but they also increased the use of more 'primitive' strategies, as the strategy of repeated addition or subtraction, and the doubling or halving strategies. This shows that there is room for change and amelioration, while it seems that students encountered many mental strategies and in general were trained in mental calculations possibly for the first time.

Comparison of student performance between the third and fourth grade

The multiplication 8 × 25 and the divisions 54 ÷ 6 and 48 ÷ 4 were given to students of both the third and fourth grades. We compare the success rates of these operations before the teaching interventions for the third and fourth grades: 8 × 25 (third grade: 47.5%, fourth grade: 72.5%) 54 ÷ 6 (third: 57.5%, fourth: 73%) and 48 × 4 (third: 55%, fourth: 68%). We remark that for all the questions the success rates increased significantly from the third to the fourth grade. It seems that students, over time, due to training and experience, improve their performance in this kind of multiplication and division.

Results from the pre- and post-test for the fifth and sixth grades

As we have already remarked, for the fifth and sixth grades the same exercises were suggested.

Success in calculations

For the fifth and sixth grades two multiplications and two divisions were given. As Table 4.6 shows, the multiplication 8 × 25 had a high success rate (86.5%), two-digit multiplication and division (50 × 46 and 450 ÷ 15) had a 71% success rate and the division 4200 ÷ 60 dropped to 58.5%. After the teaching intervention the success rates for these four operations was increased significantly in all operations, including 8 × 25, which increased to the limit of the statistical significance.

Use of strategies

A quite high proportion of students use written algorithms to carry out mental multiplication and division. Before instruction, this rate was on average 22.5%, and, surprisingly, after instruction in mental calculation, this rate, instead of dropping, increased to 29.5%. It seems that the processes of fifth- and sixth-grade students are not improved by teaching intervention; they do not use flexible strategies, on the contrary, they use more algorithmic methods. Low percentages of students (about 10%) used holistic methods in multiplications both before and after the teaching intervention. Only in the operation 50 × 46 did the rate of students that used the holistic strategies double (by 7% to 14.5%) after teaching intervention. For this operation, the strategy of the simultaneous doubling of 50 and the halving of 46 is appropriate: 50 × 46 = 100 × 23.

Comparison among the fourth, fifth and sixth grades

Common exercises suggested for the fourth, fifth and sixth grades were the two-digit multiplications 8 × 25 and 50 × 46 and the division with a two-digit divisor 450 ÷ 15. We note that in all three operations after the experimental intervention the success rates increased significantly: 8 × 25: 72.5% → 86.5%,

TABLE 4.6 Success rates of strategies used before and after experimental teaching in the fifth and sixth grades

Operation	Number of correct answers N = 308		Written algorithm		Repeated addition or subtraction		Partitioning number strategies		Holistic or compensation strategies		Operation production (division only)	
8 × 25	266	**278**	80	**99**	38	**38**	81	**76**	35	**32**	2	**0**
	86.5%	**90.5%**	26%	**32%**	12.5%	**12.5%**	26.5%	**24.5%**	11.5%	**10.5%**	0.5%	
50 × 46	218	**247**	68	**96**	7	**19**	91	**72**	22	**45**	0	**0**
	71%	**80%**	22%	**31%**	2.5%	**6%**	29.5%	**23.5%**	7%	**14.5%**	0%	
450 ÷ 15	220	**248**	69	**77**	5	**6**	16	**17**	1	**1**	94	**109**
	71.5%	**80.5%**	22.5%	**25%**	1.5%	**2%**	5%	**5.5%**	0.5 %	**0.5%**	30.5%	**35.5%**
4200 ÷ 60	180	**228**	60	**89**	3	**1**	22	**17**	1	**1**	75	**99**
	58.5%	**74%**	19.5%	**29%**	1%	**0.5%**	7%	**5.5%**	0.5%	**0.5%**	24.5%	**32%**

50 × 46: 53.5% → 71% and 450 ÷ 15: 53% → 71.5%. We see that here, too, over time, from the fourth to fifth and sixth grades, students have improved their performance in these kind of multiplications and divisions.

4.3.2.3. Studies of the strategies of high school students
Study of Cypriot high school students

A survey was conducted among 198 Cypriots students of the second and third years of high school. The aim of this study was to examine the abilities of students and the strategies they use when they mentally calculate multi-digit multiplication. A booklet containing eight multiplication exercises was submitted to the students, in which they were asked to reply, after coming up with an answer mentally, and then write the procedure that they used to find the answer.

From the data in the table below, a first observation that we can make is that, although it was emphasised that students were not to use a written algorithm for the multiplication, but to calculate mentally for all the exercises, about 20% of students calculated using the written algorithm. We also observe that very few students calculated mentally using the addition strategy, in relation to the previous research. This can be interpreted by the fact that, apart from that the numbers are large and not suitable for repeated addition, students are now older and use other methods to multiply. They have memorised more multiplication facts and additionally know the written algorithm for multiplication.

TABLE 4.7 Success rates and strategies used by students in the second and third year of high school in two-digit multiplications

Exercises	Success	Addition strategies	Partitioning not based on place value	Partitioning based on place value	Compensating or holistic strategies	Vertical written algorithm
5 × 12	192	13	7	91	9	31
	97%	6.5%	3.5%	46%	4.5%	15.5%
9 × 16	169	8	6	99	14	37
	85.5%	4%	3%	50%	7%	18.5%
12 × 14	159	1	10	107	8	43
	(80.5%)	0.5%	5%	54%	4%	21.5%
8 × 99	160	1	9	63	53	41
	81%	0.5%	4.5%	32%	27%	20.5%
25 × 480	129		2	85	24	44
	65%	–	1%	43%	12%	22%
103 × 15	137	2	1	109	9	41
	69%	1%	0.5%	55%	4.5%	20.5%
15 × 48	130	1	2	91	24	39
	65.5%	0.5%	1%	46%	12%	19.5%
50 × 46	151		4	86	29	41
	76.5%	–	2%	43.5%	14.5%	20.5%

The mental strategy used most by these students was the strategy of partitioning the number into place values. Approximately 50% of students used this strategy to mentally calculate multiplication.

According to the percentages in Table 4.7, we observe that students did not usually use holistic or compensating strategies. In the multiplication 8 × 99, the highest rate of use (27%) of this strategy was presented. Forty-one students (20.5% of students), in order to multiply 8 × 99, rounded 99 to 100 and calculated 8 × 100 = 800, 800 − 8 = 792. There were only two students (1%) who rounded the 8 and calculated 10 × 99 = 990, 990 − 99 = 891, 891 − 99 = 792.

Apart from the operation 8 × 99, other multiplications that are offered for holistic strategies are:

- Multiplication 9 × 16 → 10 × 16 − 16, where only 13 students (6.5%) applied this strategy.
- For multiplication 25 × 480 → 25 × 500 − 25 × 20 = 12,500 − 500 = 12,000, only 8 students (4%) applied this strategy. Seven students (3.5%) applied 480 × 100 = 48,000, 48000 ÷ 4 = 12,000 or 50 × 480 = 24,000, 24,000 ÷ 2 = 12,000.
- Multiplication 15 × 48 → 15 × 50 = 750, 2 × 15 = 30, 750 − 30 = 720, where 15 students (7.5%) applied this strategy.
- Multiplication 50 × 46. Here 15 students (7.5%) rounded the number 46 → 50 × 50 = 2,500, 50 × 4 = 200, 2,500 − 200 = 2,300 and 13 students (6.5%) rounded the 50 → 46 × 100 = 4,600, 4,600 ÷ 2 = 2,300.

The fact that only a few students used holistic strategies demonstrates that students do not have flexibility in mental calculations. This indicates that they have not practised mental calculations enough using numerous strategies of computation to enable them to choose the quickest and most effective among them.

4.4. Greek multiplication: introduction to a standard multiplication algorithm

In Section 2.2.4.4, which referred to mental calculations and standard written algorithms, calculations in columns (column calculations) were highlighted. The multiplication of 23 × 15, for example, is carried out in columns, as below:

```
        2 3                                    2 3
      × 1 5                                  × 1 5
      -----                                  -----
      2 0 0                                  1 1 5
        3 0               ⟶                + 2 3 0
      1 0 0                                  -----
    +   1 5                                  3 4 5
      -----
      3 4 5                              Standard written
   Column calculation                        algorithm
```

This way of calculating the multiplication in columns, as was already mentioned, constitutes the connecting link between mental and standard written algorithmic calculations. In column multiplication, numbers are divided into units, tens, etc. and the distributive property of multiplication is applied with respect to addition: 23 × 15 = (20 + 3) × (10 + 5) = (20 × 10) + (20 × 5) + (3 × 10) + (3 × 5). The range of the digits, in order for individual multiplications to take place, is from left to right instead of right to left, as in the typical algorithm. In multiplication in columns, whole numbers are used, but not in the form of individual digits, as in standard algorithms. More specifically, multiplication in columns will be further explored, as this mode of reproduction has a historically Greek origin; for this reason it is called *Greek multiplication*.

It can be easily observed that this method closely resembles the classic algorithms used today. However *Greek multiplication* is more extensive, because all the individual cross-products between the digits of the two factors are presented. In the classical algorithm, on the other hand, the individual cross-products are generated by multiplying each digit of the multiplier by the entire multiplicand. The classical algorithm is denser and shorter, but at the same time it is quite complex to explain it to students. *Greek multiplication* seems simpler and can be used before the classic algorithm to explain the logic behind its properties.

Greek multiplication is not commonly mentioned in the literature. Nonetheless, it has been referenced by name (as *Greek multiplication*) by a few French authors (e.g., Cerquetti-Aberkane, 1992, p. 79) without an explanation of its historical origin. It is thought that this name was given to this type of multiplication because the ancient Greek mathematicians negotiated numerical problems by using geometrical shapes. Numbers were represented by lengths, the product of two numbers by the areas and the product of three numbers by volumes. The fact that in the context of mathematical terminology the power 2 is called 'square' and the power 3 'cube' is due to the geometric representation used by the Greeks in order to execute calculations with cross-products.

We will present the historical origin of Greek multiplication and its implementation of education.

4.4.1. Historical origins of Greek multiplication

The text and data in this paragraph are derived from the work of Lemonidis and Nikolantonakis (2007). Historical research was necessary in order to determine how Greek multiplication was presented in the history of mathematics. In this historical research, carried with the original texts, it was found that the Greek multiplication is described in the comments of Eutocius of Ascalon (around the fifth century AD) on the treatise of Archimedes entitled *The measurement of the circle*. Eutocius performed multiplications in order to find the square of a number. For this, he was required to multiply a number by itself each time. He separated the number into units, tens, hundreds, etc. and calculated the partial products between units of the same class, which he added in order to obtain a final result.

4.4.1.1. Biographical sketch of Eutocios of Ascalon

Very little information about the life of Eutocius has survived to the present day (Heath, 2001, II, pp. 540–1; Decorps-Foulquier, 1994). According to information in his treatises, he came from Ascalon, in the province of Syria. Relying on a relatively small amount of information, it can be assumed that *terminus post quem* for the date of his birth was around 450 AD, while *terminus post quem* for the date of his death was the year 497 AD.

4.4.1.2. The treatise The measurement of the circle *and Greek multiplication*

Eutocius produced several comments on the works of Archimedes: *On the sphere and cylinder*, *The measurement of the circle* and *On plane equilibria*, as well as a critical edition of books 1–4 of the *Conics* of Apollonius of Perga. The treatise *The measurement of the circle* consists of three propositions and has not yet reached us in its original form, the version we possess having practically lost all traces of the Doric dialect in which Archimedes wrote. Of these three propositions, the third proposition, the one in which we are interested, is the following: 'The ratio of the circumference of each circle to its diameter is less than 3 1/7 and greater than 3 10/71'. This proposition contains a numerical approximation of π. This method is equivalent to an approximate calculation of the perimeter of two regular polygons with 96 sides, of which the one is circumscribed and the other is inscribed in the circle. The calculation starts with an upper and a lower limit of the value $\sqrt{3}$. Archimedes, in the relation $265 \div 153 < \sqrt{3} < 1351 \div 780$ does not show the calculations he executed, but presents it without any comment.

In his comments on the third proposition, Eutocius noted that in this, it was constantly required to find the square root of a given number, but it was impossible to find an accurate value of this size for a non-square number; whereas the sum of a number and a fraction multiplied by itself does not give a pure number, but a fraction. Concerning methods for finding the approximate values of the square root of a given number, he refers to the *Metrics* of Heron, to Pappus, to Theon and to other commentators on the *Almagest* of Claudis Ptolemy, highlighting that research on this particular question by him was not necessary. Eutocius used Archimedes' treatise as a manual, analysing every proposition of the demonstration, explaining the reasons for its truth and also performing a series of operations that did not exist in the original demonstration of Archimedes (Mugler, 1972, pp. 142–163). Eutocius, through encountering concrete examples, teaches us the methods of the numerical calculation of the Greeks and thus shows us the so-called 'Greek multiplication'.

Overall, Eutocius in his comments about Archimedes' treatise *The measurement of the circle* analysed the following multiplications, always in square numbers (i.e., the product of a number multiplied by itself):

$$306, 153, 265, 571, 591\frac{1}{8}, 1162\frac{1}{8}, 1172\frac{1}{8}, 2334\frac{1}{4}, 2339\frac{1}{4}, 1560, 780, 1351,$$

$$2911, 3013\frac{1}{2}\frac{1}{4}, 1823, 240, 1838\frac{1}{9}\frac{1}{11}, 1007, 66, 1009\frac{1}{6}, 2016\frac{1}{6}, 2017\frac{1}{4}.$$

150 Multiplication and division

The examples of the three multiplications shown below, using modern symbolism of one fractional and two integer numbers, were written by Eutocius in the ancient text.

The first example is the multiplication 66 × 66:

$$\overline{\eta} \; \mathrm{K}\Gamma \; \overline{\xi\varsigma}$$
$$\dot{\epsilon}\pi\dot{\iota} \; \overline{\xi\varsigma}$$
$$,\mathrm{YX} \; \overline{\tau\xi}$$
$$\overline{\tau\xi} \; \overline{\lambda\varsigma}$$
$$\dot{\delta}\mu o\hat{\upsilon} \; ,\overline{\delta\tau\upsilon\varsigma}$$

In modern symbolism, this multiplication takes the following form:

		66
	×	66
3,600		360
360		36
Total		4,356

The mathematical analysis of these operations can be described as follows:

(6 Tens + 6 Units) × (6 Tens + 6 Units) =
36 Hundreds + 36 Tens +
36 Tens + 36 Units

The second example includes the multiplication 1351 × 1351:

$$,\overline{\alpha\tau\nu\alpha}$$
$$\dot{\epsilon}\pi\dot{\iota} \; ,\overline{\alpha\tau\nu\alpha}$$
$$\begin{array}{c}\rho\lambda\epsilon\\ \mathrm{MMM} \, \overline{,\alpha}\end{array}$$
$$\begin{array}{c}\lambda\theta\alpha\\ \mathrm{MMM} \, \overline{,\epsilon\tau}\end{array}$$
$$\begin{array}{c}\epsilon\alpha\\ \mathrm{MM} \, \overline{,\epsilon} \, \overline{,\beta\phi\nu}\end{array}$$
$$,\overline{\alpha\tau\nu\alpha}$$
$$\dot{\delta}\mu o\hat{\upsilon} \quad \begin{array}{c}\rho\pi\beta\\ \mathrm{M} \, \overline{'\epsilon\sigma\alpha}\end{array}$$

Multiplication and division **151**

In modern symbolism, the multiplication takes the following form:

				1,351
			×	1,351
1,000,000	300,000	50,000		1,000
300,000	90,000	15,000		300
50,000	15,000	2,500		50
1,000	300	50		1
		Total		1,825,201

The mathematical analysis of these operations can be described as follows:

(1 Thousand + 3 Hundreds + 5 Tens + 1 Unit) × (1 Thousand + 3 Hundreds + 5 Tens + 1 Unit) =

1 Million + 3 Hundred Thousands + 5 Myriads + 1 Thousand

3 Hundred Thousands + 9 Myriads + 15 Thousands + 3 Hundreds

5 Myriads + 15 Thousands + 25 Hundreds + 5 Tens

1 Thousand + 3 Hundreds + 5 Tens + 1 Unit

The third example includes a fractional number.

$$,\overline{γιγ}\; L'\; δ'$$
$$ἐπὶ\; ,\overline{γιγ}\; L'\; δ'$$
$$\overset{\lambda γ}{MM}\; ,\overline{θ}\; \overline{,αφ}\; \overline{ψν}$$
$$\overset{γ}{M}\; \overline{ρλε}\; \overline{β}\; L'$$
$$,\overline{θλθ}\; \overline{α}\; L'\; L'\; δ'$$
$$,\overline{αφ}\; \overline{ε}\; \overline{α}\; L'\; δ'\; η'$$
$$,\overline{ψν}\; \overline{β}\; L'\; L'\; δ'\; η'\; ιϛ'$$
$$ὁμοῦ\; \overset{\lambda η}{M}\; \overline{,βχπθ}\; ιϛ'$$

152 Multiplication and division

				$3,013\frac{1}{2}\frac{1}{4} = 3,013\frac{3}{4}$
			×	$3,013\frac{1}{2}\frac{1}{4}$
9,000,000	30,000	9,000	1,500	750
30,000	100	30	5	$2\frac{1}{2}$
9,000	30	9	$1\frac{1}{2}$	$\frac{1}{2}\frac{1}{4}$
1,500	5	$1\frac{1}{2}$	$\frac{1}{4}$	$\frac{1}{8}$
750	$2\frac{1}{2}$	$\frac{1}{2}\frac{1}{4}$	$\frac{1}{8}$	$\frac{1}{16}$
			Total	9,082,689

In modern symbolism, the multiplication takes the following form:
The mathematical analysis of the operations above can be described as follows:

$(3 \text{ Thousands} + 1 \text{ Ten} + 3 \text{ Units} + \frac{1}{2} + \frac{1}{4}) \times (3 \text{ Thousands} + 1 \text{ Ten} + 3 \text{ Units} + \frac{1}{2} + \frac{1}{4}) =$

9 Hundred Myriads + 30 Myriads + 9 Thousands + 15 Hundreds + 75 Tens +

3 Myriads + 1 Hundred + 3 Tens + 5 Units + 2 Units + $\frac{1}{2}$ +

9 Thousands + 3 Tens + 9 Units + 1 Unit + $\frac{1}{2}$ + $\frac{1}{2}$ + $\frac{1}{4}$

15 Hundreds + 5 Units + 1 Unit + $\frac{1}{2}$ + $\frac{1}{4}$ + $\frac{1}{8}$

75 Tens + 2 Units + $\frac{1}{2}$ + $\frac{1}{2}$ + $\frac{1}{4}$ + $\frac{1}{8}$ + $\frac{1}{16}$

4.4.2. Implementation of Greek multiplication in education

In the historical analysis, it was stated that Greek multiplication is a very old multiplication algorithm, appearing in the fifth AD century texts of Eutocius, who came through the tradition of Greek mathematics. The algorithm of Greek multiplication greatly resembles the current classical algorithm, which has an Indian and Arabian origin. Both styles use the same mathematical properties, namely the distributive property of multiplication, regarding in addition, and cumulative analysis of multiplication factors, based on the place value of the decimal number system. According to this as well as the fact that the Indian–Arabic multiplication algorithm is historically posterior, it could be thought that Greek multiplication is an original version and progenitor of the modern classical algorithm.

The calculation technique of Greek multiplication can be considered as suitable for teaching. This technique could be a bridge between mental calculation and written standard multiplication algorithm. Moreover, this can be confirmed by its use in many innovative curricula, for example those of countries like the Netherlands (multiplication by calculating columns, Section 2.2.4.4), England, etc. It can be said that it is easy and compatible with the human way of thinking in calculations. This is demonstrated by the fact that it is a method invented by students (informal method) in order to calculate multiplications before being taught any multiplication methods (see Section 4.3.2.1). In this method, the terms of multiplication are analysed into sum powers of ten (units, tens, etc.). This way, during calculations, whole quantities are used of which we know the value, rather than individual digits of the value to which is made no reference, as in the classic algorithm. This feature of Greek multiplication makes it appropriate and useful for the educational process, just because it is illustrative in terms of value and in the way the partial result of multiplication arises.

In the new Greek primary school books of third-grade mathematics, the Greek multiplication is presented with a short historical note (Lemonidis et al., 2006, pp. 72–77). The introduction of Greek multiplication in education is in accordance with the principles of the Mathematics of Nature and Life. The contents and modes used for the operations are intended to be fun for children and in accordance with their prior knowledge. These contents also refer to the nature, culture and history of mathematics (Nikolantonakis & Lemonidis, 2008).

In a third-grade textbook, Greek multiplication is used as a preliminary stage for the introduction of the standard algorithm used today. More specifically, the first phase is connected with students' multiplication problems with multi-digit numbers, in order to develop informal methods of calculation. Greek multiplication is inserted through a geometric frame, i.e., in situations such as measuring surfaces on graph paper. Students measure the surfaces of squares and rectangles using one small square as a unit of measurement. Then, for an easier and quicker calculation of surfaces, they use the 2 × 2 multiplication table. Hence, they end up being able to perform the Greek multiplication with the help of the table. Moreover, the standard algorithm can be explained based on Greek multiplication, in order for students to understand how individual products and the place value of the digits of the factors of the multiplication are produced.

This historic multiplication algorithm can possibly be applied to other levels of education related to the use of history of mathematics in teaching. Last but not least, it can also be used for teaching children who have learning difficulties in mathematics.

References

Anghileri, J. (1989). An investigation of young children's understanding of multiplication. *Educational Studies in Mathematics, 20*, 367–385.

Ashcraft, M. H. (1995). Cognitive psychology and simple arithmetic: A review and summary of new directions. *Mathematical Cognition, 1*, 3–34.

Ashcraft, M. H. & Christy, K. S. (1995). The frequency of arithmetic facts in elementary texts: Addition and multiplication in grades 1–6. *Journal for Research in Mathematics Education, 26*, 396–421.

Askew, M. & Wiliam, D. (1995). *Recent research in mathematics education 5–16.* London: HMSO.

Baek J.-M. (1998). Children's invented algorithms for multidigit multiplication problems. In L. J. Morrow and M. J. Kenney (Eds.), *The teaching and learning of algorithms in school mathematics, 1998 yearbook* (pp. 151–156). Reston, Va.: National Council of Teachers of Mathematics.

Baroody, A. J. (1999). The roles of estimation and the commutativity principle in the development of third graders' mental multiplication. *Journal of Experimental Child Psychology, 74,* 157–193.

Brown, S. (1992). Second-grade children's understanding of the division process. *School Science and Mathematics, 92*(2), 92–95.

Bryant, P. (1997). Mathematical understanding in the nursery school years. In T. Nunes & P. Bryant (Eds.), *Learning and teaching mathematics: An international perspective* (pp. 53–67). East Sussex, UK: Psychology Press.

Bull, R., Johnston, R. S. & Roy, J. A. (1999). Exploring the roles of the visual-spatial sketchpad and central executive in children's arithmetical skills: Views from cognition and developmental neuropsychology. *Developmental Neuropsychology, 15,* 421–442.

Campbell, J. I. D. (1994). Architectures for numerical cognition. *Cognition, 53,* 1–44.

Campbell, J. I. D. (1995). Mechanisms of simple addition and multiplication: A modified network-interference theory and simulation. *Mathematical Cognition, 1,* 121–164.

Campbell, J. I. D. (1997). On the relationship between skilled performance of simple division and multiplication. *Journal of Experimental Psychology: Learning, Memory, and Cognition, 23,* 1140–1159.

Campbell, J. I. D. (1999). Division by multiplication. *Memory & Cognition, 27,* 791–802.

Campbell, J. I. D., Fuchs-Lacelle, S. & Phenix, T. S. (2006). Identical elements model of arithmetic memory: Extension to addition and subtraction. *Memory, & Cognition, 34,* 633–647.

Campbell, J. I. D. & Graham, D. J. (1985). Mental multiplication skill: Structure, process, and acquisition. *Canadian Journal of Psychology, 39,* 338–366.

Campbell, J. I. D. & Robert, N. D. (2008). Bidirectional associations in multiplication memory: Conditions of negative and positive transfer. *Journal of Experimental Psychology: Learning, Memory, and Cognition, 34,* 54–555.

Campbell, J. I. D. & Xue, Q. (2001). Cognitive arithmetic across cultures. *Journal of Experimental Psychology: General, 130,* 299–315.

Carpenter, T. P., Ansell, E., Franke, M. L., Fennema, E. & Weisbeck, L. (1993). Models of problem solving: A study of kindergarten children's problem-solving processes. *Journal for Research in Mathematics Education, 24,* 428–441.

Cerquetti-Aberkane, F. (1992). *Enseigner les mathématiques à l'école.* Paris: Hachette.

Clark, F. B. & Kamii, C. (1996). Identification of multiplicative thinking in children in Grades 1–5. *Journal for Research in Mathematics Education, 27*(1), 41–51.

Cohen, L. & Dehaene, S. (1994). Amnesia for arithmetical facts: A single case study. *Brain and Language, 47,* 214–232.

Cohen, L. & Dehaene, S. (2000). Calculating without reading: Unsuspected residual abilities in pure alexia. *Cognitive Neuropsychology, 17,* 563–583.

Cooney, J. B., Swanson, H. L. & Ladd, S. F. (1988). Acquisition of mental multiplication skill: Evidence for the transition between counting and retrieval strategies. *Cognition & Instruction, 5,* 323–345.

Correa, J. (1994). *Young children's understanding of the division concept* (unpublished PhD thesis). University of Oxford.

Correa, J., Nunes, T. & Bryant, P. (1998). Young children's understanding of division: The relationship between division terms in a noncomputational task. *Journal of Educational Psychology*, 90(2), 321–329.

Dagenbach, D. & McCloskey, M. (1992). The organization of arithmetic facts in memory: Evidence from a brain-damaged patient. *Brain and Cognition*, 20, 345–366.

De Brauwer, J. & Fias, W. (2011). The representation of multiplication and division facts in memory: Evidence for cross-operation transfer without mediation. *Experimental Psychology*, 58(4), 312–323.

De Brauwer, J. & Fias, W. (2009). A longitudinal study of children's performance on simple multiplication and division problems. *Developmental Psychology*, 45, 1480–1496.

De Brauwer, J., Verguts, T. & Fias, W. (2006). The representation of multiplication facts: Developmental changes in the problem size, five, and tie effects. *Journal of Experimental Child Psychology*, 94, 43–56.

Decorps-Foulquier M. (1994). *Les Coniques d'Apollonios de Perge. Histoire du texte des Livres I–IV. Edition critique et traduction du Livre I.* [The Conics of Apollonius of Perga. History of the text of Books I-IV. Critical edition and translation of Book I.] Thèse de Doctorat d'Etat, Vol. 1, Fascicule 1. Clemont-Ferrand, France.

Downton, A. (2008). Links between children's understanding of multiplication and solution strategies for division. In M. Goos, R. Brown & K. Makar (Eds.), *Navigating currents and charting directions (Proceedings of the 31st annual conference of the Mathematics Education Research Group of Australasia).* Brisbane: MERGA.

Fischbein, E., Deri M., Nello, M-S. & Marino, M-S. (1985). The role of implicit models in solving verbal problems in multiplication and division. *Journal for Research in Mathematics Education*, 16(1), 3–17.

French, D. (2005). Double, Double, Double. *Mathematics in School*, 34(5), 8–9.

Frydman, O. & Bryant, P. (1988). Sharing and the understanding of number equivalence by young children. *Cognitive Development*, 3, 323–339.

Galfano, G., Russconi, E. & Umilta, C. (2003). Automatic activation of multiplication facts: Evidence from the nodes adjacent to the product. *The Quarterly Journal of Experimental Psychology*, 56A(1), 31–61.

Geary, D. C. (1990). A componential analysis of an early learning deficit in mathematics. *Journal of Experimental Child Psychology*, 49, 363–383.

Girelli, L., Delazer, M., Semenza, C. & Denes, G. (1996). The representation of arithmetical facts: Evidence from two rehabilitation studies. *Cortex*, 32, 49–66.

Greer, B. (1992). Multiplication and division as models of situations. In D. A. Grouws (Ed.), *Handbook of research on mathematics teaching and learning* (pp. 276–295). New York: Macmillan.

Haylock, D. & Cockburn, A. (1997). *Understanding mathematics in the lower primary years.* London: Paul Chapman Publishing.

Heath T. L. (2001). *Ιστορία των Ελληνικών Μαθηματικών, Από το Θαλή στον Ευκλείδη, Τόμος Ι.* [History of Greek mathematics. From Thales to Euclid, Volume I.] Athens: Κ.Ε.ΕΠ.ΕΚ.

Heirdsfield, A. M., Cooper, T. J., Mulligan, J. & Irons, C. J. (1999). Children's mental multiplication and division strategies. In O. Zaslavsky (Ed.), *Proceedings of the 23rd Psychology of Mathematics Education Conference*, pp. 89–96, Haifa, Israel.

Hitch, G. J. (1978a). Mental arithmetic: Short-term storage and information processing in a cognitive skill. In A. M. Lesgold, J. W. Pelligrino, S. D. Fokkema & R. Glaser (Eds.), *Cognitive psychology and instruction* (pp. 331–338). New York: Plenum.

Hitch, G. J. (1978b). The role of short-term working memory in mental arithmetic. *Cognitive Psychology*, 10, 302–323.

Koshmider, J. W. & Ashcraft, M. H. (1991). The development of children's mental multiplication skills. *Journal of Experimental Child Psychology*, *51*, 53–89.

Kouba, V. L. (1989). Children's solution strategies for equivalent set multiplication and division word problems. *Journal for research in Mathematics Education*, *20*, 147–158.

LeFevre, J., Bisanz, J., Daley, K. E., Buffone, L., Greenham, S. L. & Sadesky, G. S. (1996). Multiple routes to the solution of single-digit multiplication problems. *Journal of Experimental Psychology: General*, *125*, 284–306.

LeFevre, J.-A. & Liu, J. (1997). The role of experience in numerical skill: Multiplication performance in adults from Canada and China. *Mathematical Cognition*, *3*, 31–62.

LeFevre, J.-A. & Morris, J. (1999). More on the relation between division and multiplication in simple arithmetic: Evidence for mediation of division solutions via multiplication. *Memory & Cognition*, *27*, 803–812.

Lemaire, P. & Siegler, R. S. (1995). Four aspects of strategic change: Contributions to children's learning of multiplication. *Journal of Experimental Psychology: General*, *124*, 83–97.

Lemonidis, Ch., Theodorou, E., Nikolantonakis, K., Panagakos, I. & Spanaka, A. (2006). Μαθηματικά Γ' Δημοτικού. Μαθηματικά της Φύσης και της Ζωής. Βιβλίο μαθητή. [Third-grade mathematics. Nature and Life Mathematics. Students' book.] Athens: Οργανισμός Εκδόσεων Διδακτικών Βιβλίων (ΟΑΕΔ), Υπουργείο Παιδείας.

Lemonidis, Ch. & Nikolantonakis, K. (2007). Ελληνικός πολλαπλασιασμός: Ένας άγνωστος ιστορικός αλγόριθμος κατάλληλος για τη διδασκαλία. [Greek multiplication: An unknown historical algorithm suitable for teaching.] *Σύγχρονη Εκπαίδευση*, *151*, 169–178.

Lemonidis, Ch., Tsakiridou, H, Panou, F. & Griva, H. (2014). Prospective teacher's efficiency and flexibility in prep and mental calculation of two-digit multiplications. *MENON: Journal Of Educational Research. 1st Thematic Issue*, 110–125.

Logie, R. H., Gilhooly, K. J. & Wynn, V. (1994). Counting on working memory in arithmetic problem solving. *Memory & Cognition*, *22*, 395–410.

Marton, F. & Neuman, D. (1996). Phenomenography and children's experience of division. In L. R. Steffe, P. Nosher, P. Cobb, G. A. Goldin & B. Greer (Eds.), *Theories of mathematical learning* (pp. 315–333). Mahwah, NJ: Lawrence Erlbaum Associates.

Marshall, S. P. (1995). *Schemas in problem solving*. New York: Cambridge University Press.

McCloskey, M., Aliminosa, D. & Sokol, S.M. (1991). Facts, rules and procedures in normal calculation: Evidence from multiple single-patient studies of impaired arithmetic fact retrieval. *Brain and Cognition*, *17*, 154–203.

Miller, K. (1984). Child as the measurer of all things: Measurement procedures and the development of quantitative concepts. In C. Sophian (Ed.), *Origins of cognitive skills* (pp. 193–228). Hillsdale, NJ: Lawrence Erlbaum Associates, Inc.

Miller, K., Perlmutter, M. & Keating, D. (1984). Cognitive arithmetic: Comparison of operations. *Journal of Experimental Psychology: Learning, Memory, & Cognition*, *10*, 46–60.

Mugler C. (1972). *Archimède IV, commentaires d'Eutocius et fragments*. [Archimedes IV, Eutocius' commentaries and fragments.] Paris: C.U.F.

Mulligan, J. & Mitchelmore, M. (1997). Young children's intuitive models of multiplication and division. *Journal for research in Mathematics Education*, *28*(3), 309–330.

Murray, H., Olivier, A. & Human, P. (1994). Fifth graders' multi-digit multiplication and division strategies after five years' problem-centered learning. *International Group for the Psychology of Mathematics Education*, *18*(3), 399–406.

Nesher, P. (1988). Multiplicative school word problems: Theoretical approaches and empirical findings. In J. Hiebert & M. Behr (Eds.), *Number concepts and operations in the middle grades* (pp. 19–40). Hillsdale, NJ: Erlbaum.

Nikolantonakis, K. & Lemonidis, Ch. (2008). Multiculturalism, history of mathematics and schoolbook of the third class in primary school in Greece. In J. F. Matos, P. Valero & K. Yasukawa (Eds.), *Proceeding of the Fifth International Mathematics Education and Society Conference* (pp. 398–405). Lisbon: Aalborg University.

Oliver, A., Murray, H. & Human, P. (1991). Children's solution strategies for division problems. In J. Ferrini-Mundy (Ed.), *Proceedings of the Sixteenth International Conference for the Psychology of Mathematics Education, Vol. 2* (pp. 152–159). Durham, New Hampshire: PME.

Parmar, R. S. (2003). Understanding the concept of 'division': Assessment considerations. *Exceptionality, 11*, 177–189.

Penner-Wilger, M., Leth-Steensen, C. & LeFevre, J-A. (2002). Decomposing the problem-size effect: A comparison of response time distributions across cultures. *Memory & Cognition, 30*(7), 1160–1167.

Piaget, J. (1972/1947). *The psychology of intelligence* (M. Piercy & D.E. Berltyne, trans.). Totowa, NJ: Littlefield, Adams & Co.

Reed, S. K. (1999). *Word problems: Research and curriculum reform*. Mahwah, NJ: Lawrence Erbaum.

Rickard, T. C. (2005). A revised identical model of arithmetic fact representation. *Journal of Experimental Psychology: Learning, Memory, and Cognition, 31*, 250–257.

Rickard, T. C. & Bourne, L. E., Jr. (1996). Some tests of an identical elements model of basic arithmetic skills. *Journal of Experimental Psychology: Learning, Memory, and Cognition, 22*, 1281–1295.

Rickard, T. C., Healy, A. F. & Bourne, L. E., Jr. (1994). On the cognitive structure of basic arithmetic skills: Operation, order, and symbol transfer effects. *Journal of Experimental Psychology: Learning, Memory, and Cognition, 20*, 1139–1153.

Robinson, K., Arbuthnott, K., Rose, D., McCarron, M., Globa, C. & Phonexay, S. (2005). Stability and change in children's division strategies. *Journal of Experimental Child Psychology, 93*(3), 224–238.

Sherin, B. & Fuson, K. (2005). Multiplication strategies and the appropriation of computational resources. *Journal for Research in Mathematics Education, 36*(4), 347–395.

Siegler, R. S. (1988). Strategy choice procedures and the development of multiplication skill. *Journal of Experimental Psychology: General, 117*, 258–275.

Siegler, R. S. & Shipley, C. (1995). Variation, selection, and cognitive change. In T. J. Simon & G. S. Halford (Eds.), *Developing cognitive competence: New approaches to process modeling* (pp. 31–76). Hillsdale, NJ: Lawrence Erlbaum.

Siegler, R. S. & Shrager, J. (1984). Strategy choice in addition and subtraction: How do children know what to do? In C. Sophian (Ed.), *Origins of cognitive skills* (pp. 229–293). Hillsdale, NJ: Erlbaum.

Steel, S. & Funnell, E. (2001). Learning multiplication facts: A study of children taught by discovery methods in England. *Journal of Experimental child Psychology, 79*, 37–55.

Steffe, L. P. (1988). Children's construction of number sequences and multiplying schemes. In J. Hiebert & M. Behr (Eds.), *Number concepts and operations in the middle grades, Vol. 2* (pp. 119–140). Hillsdale, NJ: Lawrence Erlbaum.

Ter Heege, H. (1985). The acquisition of basic multiplication skills. *Educational Studies in Mathematics, 16*, 375–388.

Trachilou, E., Christou, Z. & Lemonidis, Ch. (2008). Οι άτυπες στρατηγικές που χρησιμοποιούν οι μαθητές στον πολλαπλασιασμό. [Informal strategies which students use in multiplication.] *Proceedings of the 10th Pancyprian Conference in Mathematics Education and Science*, Paphos 1–3 February, 449–462.

Walker, D., Bajic, D., Mickes, L., Kwak, J. & Rickard, T. (2014). Specificity of children's arithmetic learning. *Journal of Experimental Child Psychology, 122*, 62–74.

Walker, D., Mickes, L., Bajic, D., Nailon, C. & Rickard, T. C. (2013). A test of two methods of arithmetic fluency training and implications for educational practice. *Journal of Applied Research in Memory and Cognition, 2*, 25–32.

Whalen, J., McCloskey, M., Lindemann, M. & Bouton, G. (2002). Representing arithmetic table facts in memory: Evidence from acquired impairments. *Cognitive Neuropsychology, 19*(6), 505–522.

Whetstone, T. (1998). The representation of arithmetic facts in memory: Results from retraining a brain-damaged patient. *Brain and Cognition, 36*, 290–306.

Vergnaud, G. (1981). *L'enfant, la mathématique et la réalité.* [The child, mathematics and reality.] Berne: Peter Lang.

Vergnaud, G. (1983). Multiplicative Structures. In R. Lesh & M. Landau (Eds.), *Acquisition of mathematics concepts and processes.* Academic Press, Inc (London).

Vergnaud, G. (1988). Multiplicative structures. In J. Hiebert & M. Behr (Eds.), *Number concepts in the middle grades* (pp. 141–161). Hillsdale, NJ: Lawrence Erlbaum.

5
MENTAL CALCULATION WITH RATIONAL NUMBERS

In this chapter, we will examine mental calculation with rational numbers, namely fractions, decimals and percentages (called *part–whole* numbers in the literature). First, we will refer to rational numbers to illustrate some concepts and to determine difficulties children encounter during their transition from natural to rational numbers. Then we present the different strategies that students use for rational numbers. We will present children's behaviours with regard to these kinds of calculations based on current research literature and findings of local research. We illustrate results using studies in international literature. We will also present results on the comparison of fractions. Finally, we will display research findings from teachers and studies on instruction in rational numbers.

The study of mental calculation with rational numbers is important, as students' work with mental calculations with fractions, decimals and percentages offers a deep conceptual understanding of these concepts. Students must consider these numbers through different semiotic representations and also translate and transition from one type of number to another (e.g., to consider fractions as decimals). This increases student flexibility.

In addition, proportional reasoning in secondary schools, based on part–whole numbers, has drawn attention. In the logic of quantitative numeracy, for an up-to-date school curriculum with demand for high-level mathematical knowledge (e.g., Madison & Steen, 2003), it must be acknowledged that people are led to use proportional reasoning in many social contexts and for life skills, which are frequently executed mentally and are based on fractions, decimals and percentages (Steen, 2001; Watson, 2004).

In the literature there are only a few studies on students' behaviour in mental calculation with rational numbers. Despite this, in some curricula, such as the

American Standards 2000 (NCTM, 2000, p. 220–221), there are references to mental calculation with fractions, decimals and percentages:

> Students should also develop and adapt procedures for mental calculation and computational estimation with fractions, decimals, and integers. Mental computation and estimation are also useful in many calculations involving percentages. Because these methods often require flexibility in moving from one representation to another, they are useful in deepening students' understanding of rational numbers and helping them think flexibly about these numbers.

5.1. Rational numbers and fractions

There are many findings in the research concerning fractions and rational numbers in general. In the following section, we will briefly present what rational numbers and the fractions are, concentrating on the special features that differentiate them from natural numbers. Later we will present the difficulties students encounter during the transition from natural to rational numbers.

5.1.1. What are rational numbers?

Initially, we show the relationship between the concept of a rational number and the concept of the fraction; for this reason, we will present the mathematical definitions of these concepts.

Rational numbers are defined as the set of numbers Q: = {x : x = a/b where a, b ∈ Z and b ≠ 0}; these are numbers that can be expressed in fraction form a/b, where a and b are integers and b ≠ 0. The relationship, which is defined in this way: a/b = c/d ⇔ a d = b c, is an equivalence relation in Q, because it is a reflective or reflexive, symmetrical and transitive relationship. Every element of Q, namely every rational number, constitutes an equivalence class, according to this relationship. This means that rational numbers represented as fractions a/b constitute an equivalence class [a, b]. For example, fractions 2/3, 4/6, 8/12, . . . all belong to the equivalence class of 2/3 and have the same value.

A *fraction* is defined as an ordered pair (a, b) in the form of a/b, where a, b are natural numbers with b ≠ 0, for which the following mathematical relationships are valid:

A) Relationship of equality: fractions a/b and c/d are equal if and only if a = c and b = d.
B) Relationship of order: according to which a fraction a/b < c/d if and only if ad < bc.

Therefore, a *rational number* is defined as a set of equivalent ordered pairs (a, b) or a/b, namely a set of fractions, where a and b are integers and b ≠ 0 and two pairs of fractions are equal a/b = c/d, if and only if ad = bc.

Every rational number is, therefore, a set of fractions with the same value, while every fraction is a different but equivalent representation of a rational number. Also, every integer can be considered a rational number, because it can be represented as a fraction with the unit as its denominator.

For instance, fractions 2/3, 4/6, 10/15 represent the same rational number. They constitute different expressions of the same rational number. So the fractions 2/3 and 4/6 are not identical, since 2 ≠ 4 and 3 ≠ 6, but they are equivalent. They are representations of the same rational number, equivalent expressions of the same value of a magnitude.

Therefore, rational numbers should be conceived as unique objects, unchanged by the differing symbolic representations that unite and include natural and non-natural numbers as part of the same set.

In the school context, then, the same rational number can be presented through different semiotic representations, e.g., 0.4, 2/5, 10/25, 40%, a fact that causes much difficulty and many problems for students.

Such new numbers have both similarities and differences to the natural numbers already known by students. Indeed, in the system learned up to this point, every natural number is represented by one and only one symbol. The natural numbers are associated with the absolute value of a quantity, the cardinal, and answer the question 'how many'. The unit creates the numbers by repetition of itself, and therefore constitutes a fundamental element for the formation of the numbers. The natural numbers are discrete, and, thus, each one has one and only one subsequent or precedent number. Between two successive natural numbers there is no natural number. In natural numbers, the addition and the multiplication of two numbers always gives a number larger than the terms of the operation. We can explicitly say that addition and multiplication 'make larger', while subtraction and division give a result smaller than the two terms of the operation, or 'make smaller'.

On the other side, as we have already noted, a rational number can have many semiotic representations. Rational numbers represent quantitative relations and answer the question 'how much' rather than the question 'how many'. These numbers entail the concepts of continuity and infinity: they are dense. As a consequence, no rational number has a unique subsequent number, unlike with the natural numbers, and between any two natural as well as any two rational numbers there are infinite rational numbers (for instance, see Brousseau, Brousseau & Warfield, 2007; Vamvakoussi & Vosniadou, 2004). For instance, there are no natural numbers between 1 and 2, but there are infinitely many fractional numbers. In the rational numbers, the unit can be divided in perpetuity, and new numbers will be created from this division. Addition and multiplication do not always give a greater number, they do not always 'make larger', and subtraction and division do not always 'make smaller'.

Another property of rational numbers that natural numbers do not have is that every rational nonzero number has a multiplicative inverse (e.g., the inverse of 3/5 is 5/3). This property is important for the understanding of the fractional division algorithm (we inverse the fraction–divisor and multiply) and later for the learning of algebra. Booth (1981) claims that students often have a limited understanding of inverse relations, especially in the domain of fractions, and this constitutes a hindrance while learning algebra.

These differences between natural and rational numbers lead to many difficulties for students and for adults as well.

5.1.2. Student difficulties with rational numbers

The difficulties of primary school students in learning fractions and rational numbers in general are widely known among educators and researchers, and continue in secondary and tertiary education (e.g., Bright, Behr, Post & Wachsmuth, 1988; Dufour-Janvier, Bednarz & Belanger, 1987; Kerslake, 1986; Lesh, Behr & Post, 1987; Mack, 1995; Nunes & Bryant, 2009; Stafylidou & Vosniadou, 2004; Thompson & Saldanha, 2003; Vamakoussi & Vosniadou, 2010).

In recent years, contemporary theories on numerical development have been presented. They concentrate explicitly on the development of integers, and attempt to interpret the development of the knowledge of fractions and rational numbers in general, based on the knowledge of integers, supporting the idea that learning about integers hinders the learning of rational numbers (e.g., Geary, 2006; Leslie, Gelman & Gallistel, 2008; Wynn, 2002). For instance, supporters of this *privilege theory* claim that specialised learning mechanisms make the study of integers easier than the study of fractions or other types of numbers (Gelman & Williams, 1998; Wynn, 2002). Gelman and Williams state that 'children's knowledge of natural numbers (a core domain) serves as a conceptual barrier to later learning about other numbers and their mathematical structures, for example, fractions, rations, proportions, multiplication, and division' (p. 618). Their argument is that the study of fractions is hindered, while the study of integers is assisted, because of the disposition of children to assume that every number has only one subsequent, that sets can be counted with one-to-one correspondence between numbers and objects, and that the last number counted represents the cardinality of the set.

Developmental theories share similar theses: for instance, Geary (2006) declares that integers are biologically primordial, while fractions and other types of numbers are biologically secondary. His approach, like previous approaches, considers that the limitations and the preconceptions that make the learning of integers easier (e.g., counting leads to a unique cardinal number) constitute a hindrance for the learning of fractions.

Theories of *conceptual change* (Ni & Zhou, 2005; Vamvakoussi & Vosniadou, 2004; Vosniadou, Vamvakoussi & Skopeiliti, 2008) emphasise the development of the knowledge of fractions rather than the privilege and developmental theories. These theories, similar to the others, emphasise the differences between learning integers and learning fractional numbers and highlight the way 'preconceptions of integers' hinder the learning of fractions.

Studies demonstrate that students' prior knowledge and experience affect the manipulation of symbols of rational numbers in a negative way. It has been observed that students manipulate the symbol a/b as two independent natural numbers or concentrate on an additive rather than multiplicative expression of the relationship between a and b. Additionally, it was found that for the child, when the decimal part has more digits, the number is larger or, when the terms of the fractions are large numbers, the fraction is large as well (Kerslake, 1986; Kieren, 1993; Moskal & Magone, 2000; Lamon, 1999; Moss, 2005; Stafylidou & Vosniadou, 2004).

In fractions, it is the relation between the two numbers that represents the quantity, not their independent values. Stafylidou and Vosniadou (2004) analysed the understanding of fractions among Greek students and found that most students between 11 and 13 years old do not seem able to manipulate the written representation of a fraction as a multiplicative relation between the numerator and the denominator. Only 20% of 11-year-olds, 37% of 12-year-olds and 48% of 13-year-olds employ this kind of interpretation for fractions.

Many younger students (about 38% of 10-year-olds in the fifth grade) seem to manipulate numerators and denominators as independent numbers, while others (about 20%) are able to conceive fractions as a part–whole relation, but many students (22%) cannot give a clear explanation about how to interpret a numerator and denominator.

Another issue, noted in students but also dealt with among in-service teachers, is that students manipulate fractions and decimals as different types of numbers and not as alternative representations of the same numbers (e.g., Khoury & Zazkis, 1994; O'Connor, 2001; Vergaud, 1997).

Markovits and Sowder (1991), in one of their studies, find that the majority of secondary students have difficulties understanding the relationship between decimal and fractional numbers. These students either order numbers presented both in fraction and in decimal form with difficulty, order them separately or state that ordering is not possible.

The confusion of a number with its representations is also observed in the case of equivalent fractions. The difficulty of questions of equivalent fractions is differentiated depending on the suggested situations. Kerslake (1986) notes that when students are given charts with figures divided into different numbers, and are asked to compare two fractions, they can complete this task relatively easily, because a conceptual comparison can be made. If students are given a chart with six or nine parts, and are asked to indicate 2/3 of the figure, a high percentage of them fail to detect the equivalent fractions 4/6 and 6/9. Hart, Brown, Kerslake, Kuchermann, and Ruddock (1985), with a sample of 55 students from 11 to 13 years old, found that about 60% of 11-year-olds and 12-year-olds and about 65% of 12-year-olds and 13-year-olds are able to answer these questions. Nunes, Bryant, Pretzlik and Hurry (2006) submitted the same problems to a sample of 130 students of third and fourth grade of primary school (average ages of 8.6 and 9.6, respectively). A success of 28% was marked among third graders and 49% among fourth graders. This low success rate could not be explained by the lack of comprehension of the fraction 2/3, since 78% of students in third grade and 91% of fourth graders could shade the 2/3 of the figure correctly when the chart was divided into three parts.

For a long time, research in mathematics education has dealt with proportional reasoning, which is associated with the use of fractions, decimals and percentages. Thompson and Saldanha (2003) analyse the understanding of fractions in their study. In Noelting's papers (1980a, 1980b) the connections between proportional reasoning and ratios were clarified, as well as their following links with the concepts of probabilities (Shaughnessy, 2003; Watson, Collis & Moritz, 1997).

Much research has examined the construction of conceptual understanding associated with proportional reasoning in contexts that have multiplicative rather than additive features (Cobb, 1999; Harel & Confrey, 1994; Thompson & Saldanha, 2003). Difficulties related to decimal representations are attributed to a lack of proportional reasoning: either the decimal representation stands by itself (e.g., Stacey & Steinle, 1998), or it is combined with fractions (e.g., Watson, Collis & Campbell, 1995).

Percentages, although they are closely related to fractions and to decimals, have not been extensively researched (e.g., Lembke & Reys, 1994). Dole (1999) suggested the use of a proportional number line as a suitable strategy for percentages, however not in relation with mental computation.

5.2. Strategies in mental calculation with rational numbers

There have been several studies on strategies used by students in mental calculations with rational numbers (e.g., Caney & Watson, 2003; Callingham & Watson, 2008; Clarke & Roche, 2009; Lemonidis & Kaiafa, 2014a, 2014b; Post, Cramer, Behr, Lesh & Harel, 1993; Yang, Reys & Reys, 2009).

McIntosh, De Nardi and Swan (1994) believe that strategies can be separated into *instrumental* and *conceptual*. This division by McIntosh et al. is based on Skemp's (1976) terms of *instrumental* and *relational* understanding; McIntosh et al. replace the term *relational* by the term conceptual (see Section 1.2.2).

Caney and Watson (2003) distinguish two large categories of *instrumental* and *conceptual* strategies for mental operations with rational numbers. These strategies are called *instrumental* or *procedural* when students use the techniques learned by heart and not accompanied by explanations with a conceptual understanding of the process. The *conceptual* understanding shows that they understand the relationships between key structures underlying the numbers and operations. It is possible, of course, that students demonstrate conceptual understanding of a process they use, which is described as a *mixed* strategy.

Yang and colleagues have coded subjects' strategies as *number-sense* and *rule-based* (Yang, 2003, 2005, 2007; Yang et al., 2009). We adopt the terms number-sense and rule-based strategies.

Their criterion for distinguishing a strategy as based on number sense was whether one or more components of number sense are evident in the solution process (Yang, 2003, 2005, 2007). Some examples of number-sense strategies are (a) the conversion of a fraction or a percentage to a decimal before operating, (b) the schematic representation of fractions (Caney & Watson, 2003) and

(c) residual thinking (Behr, Wachsmuth, Post & Lesh, 1984) wherein the fraction with the smaller residual is taken to be the bigger fraction. On the other hand, rule-based strategies are based on memorising rules that are not necessarily linked to deep conceptual understanding.

5.2.1. Strategies to compare fractions

Research studies demonstrate that strategies used by students in comparing fractions have not been explicitly taught (Clarke & Roche, 2009; Post, Behr & Lesh, 1986). Such strategies are *residual thinking*, *benchmarks* and the *transitive property*.

Residual thinking strategy refers to the amount needed to reach the whole. For example, in the comparison of fractions 3/4 and 5/6 a student states that the fraction of 3/4 needs 1/4 to complete the unit, while the fraction 5/6 needs 1/6 to complete the unit. Thus, he claims that the fraction 5/6 is greater because its residual distance from the unit is less.

The *benchmark* strategy refers to a student's comparison between two fractions using a reference to a third fraction, often 1/2 and sometimes 1 or 0. A student who uses this strategy claims that the fraction 5/6 is greater than 3/8, because 5/6 is bigger than 1/2 and the 3/8 is smaller. Post et al. (1986) indicate that this *benchmark* strategy is a *transitive* strategy, because the transitive property is used in regard to an external value such as 1/2 or 1 (5/6 > 1/2, 1/2 > 3/8, → 5/6 > 3/8).

Apart from the above, other strategies that originate in instruction are: *conversion to a common denominator*, in which students make equivalent fractions in order to compare, or *using equivalent fractions* (e.g., 2/4 and 4/8).

5.2.2. Examples of number sense and rule-based strategies

In Table 5.1 we present some examples of number-sense and rule-based strategies from the research of Lemonidis and Kaiafa (2014b).

5.2.3. Strategies used in mental calculations with fractions

Lemonidis and Kaiafa (2014b), found that students in the fifth grade use mental representations of objects to calculate the operation 1 − 1/4 (see Table 5.1). A student writes down a characteristic way of thought in this operation: 'I thought that 1 [the unit] was a pizza that we separated into four equal parts, and we took one part of it and three parts remained, which is 3/4'.

On the contrary, other students − in alignment with the majority of Greek students, as we will see below − use rules to answer; for instance, in the operation 1 − 1/4 they convert the numbers into equivalent fractions and find the answer, or in exercise 1/2 ÷ 1/4 they reverse the second fractions and multiply. In this case, the answer is described as *rule-based strategy*. On the other hand, a fifth-grade student gives a conceptual answer for the exercise 1/2 ÷ 1/4: '*1/4 is half of 1/2, so 1/2 ÷ 1/4 gives us two*'.

TABLE 5.1 Strategies used by students (from Lemonidis & Kaiafa, 2014b)

Strategy	Operation	Description
Execution of written algorithm (Rule-based strategy)	1 − 1/4	Converting to fractions with same denominator and subtracting.
	1/2:1/4	Reversing the second fraction and multiplying.
	Comparison of 3/7 and 5/8	Converting to fractions with same denominator and comparing numerators.
	90% of 40	90/100 × 40 = 3600/100 = 36.
Conversion of fraction or percentage to decimal (Number-sense strategy)	1 − 1/4	1/4 = 0.25, so 1 − 1/4 = 1 − 0.25 = 0.75.
	1/2:1/4	1/2 equals 0.5, 1/4 is 0.25. 0.25 goes two times into 0.5 or 0.5 is twice the 0.25.
	Comparison of 3/7 and 5/8	3/7 = 0.4 and 5/8 = 0.6. Therefore, 5/8 > 3/7.
	90% of 40	90:100 = 0.9 and 0.9 × 40 = 36.
Use of mental representations (Number-sense strategy)	1 − 1/4	I see 1 as a whole pizza or as a clock with four quarters. I take away 1/4, so 3/4 are left.
	Comparison of 3/7 and 5/8	Use of number line or of pizza as a mental representation.
Benchmarks (Number-sense strategy)	1/2:1/4	1/2 is half, 1/4 goes 2 times into half.
	Comparison of 3/7 and 5/8	5/8 is bigger than 1/2. 3/7 is smaller than 1/2. Hence, 5/8 is bigger.
Residual thinking (Number-sense strategy)	Comparison of 3/7 and 5/8	3/7 still needs 4/7 to complete a whole (7/7). 5/8 needs 3/8 to become a whole (8/8). Since 4/7 > 3/8 it's 5/8 > 3/7.
Reduction to the unit (Number-sense strategy)	90% of 40	1% of 40 is 40:100 = 0.4 Therefore, 90 × 0.4 = 36.
Benchmarks (Number-sense strategy)	90% of 40	10% of 40 is 4. 90% of 40 is 100% − 10%, so 90% of 40 is 40 − 4 = 36.
Conversion to equivalent fractions (Rule-based strategy)	1/2:1/4	Converting to fractions with the same denominator and dividing the numerators 1/2: 1/4 = 2/4: 1/4 = 2.
Creation of a composite fraction (Rule-based strategy)	1/2:1/4	Creating a composite fraction and then, multiplying the edge terms to find the numerator and the middle terms to find the denominator.

5.2.4. Strategies used in mental calculations with decimals

Caney and Watson (2003, p. 10) find that in mental operations with decimals, most students' responses were described as *instrumental* (using a *rule-based strategy*), contrary to operations with fractions and percentages. Callingham and Watson (2008, p. 96) highlight that the strategies of many students in the addition and subtraction of decimal numbers include the use of *whole number analogues*. They mention the characteristic answer of a sixth-grade student for the operation four point five minus three point three, in which the student says that he subtracts 3 from 4 and 0.3 from 0.5 to find the result 1.2. The writers indicate that this student uses the separation strategy in accordance with place value. It seems that he has an implicit conceptual understanding of the numbers contained in the operation. This strategy is common and is observed in different grades.

Callingham and Watson (2008) report that the strategies are classified as *instrumental (rule-based)* when students use *whole number analogues* and use procedures based on rules. For instance, in the exercise 0.5 + 0.75, a seventh-grade student uses whole numbers and a simulation of the written algorithm. He converts 0.5 to 50, 0.75 to 75, sums, finds 125 and then places the decimal point to find 1.25. Callingham and Watson (p. 97) state that, regardless of the students' facility with calculations with decimals, the fact that students cannot provide appropriate explanations of the position of the decimal point and the digit's place value shows that their knowledge of decimals is not very stable.

5.2.5. Strategies used in mental calculations with percentages

Caney and Watson (2003, p. 10) find that all responses of Australian students in exercises with percentages, apart from one, show number-sense strategies are used and that there is a conceptual understanding of numbers. On the contrary, most Greek students use rule-based strategies, i.e., rules to calculate percentages (Lemonidis & Kaiafa, 2014b). This difference in students' behaviours is certainly due to the way percentages are taught. In Australia, percentages are introduced with mental calculation of significant percentages, as 50%, 25%, 10%, 5%, etc., which in turn are connected with fractions, while in Greece percentages are introduced with written rules of calculation.

Lemonidis and Kaiafa (2014b), in their research on sixth-grade students, find that for the calculation of 90% of 40, the majority of students use rule-based strategies, i.e., the rule of finding rates (90/100 × 40 = 3600/100 = 36) or the proportions algorithm (90 out of 100, x out of 40). Fewer students convert the percentage to a decimal number (90 ÷ 100 = 0.9 and 0.9 × 40 = 36), which is considered a number-sense strategy. Another number-sense strategy, used by even fewer pupils, was to find 90% in regard to 10%, i.e., the following calculation: 10% of 40 is 4, 90% is 100% − 10%, so 90% of 40 is 40 − 4 = 36. Here, that is, students use known numerical facts to calculate 10% of 40, which is 4.

5.3. Research results on mental calculations with rational numbers

Research studies on mental calculation with fractions, decimals and percentages have been conducted in Australia (Callingham & Watson, 2004, 2008; Caney & Watson, 2003). The cited research was all conducted in six schools in Tasmania, Australia, from 2000 to 2004, with a sample of 5,535 students in Grades 3 to 10. In these studies, mental calculations with integers and rational numbers were examined, and eight levels of behaviour were identified in total. Mental calculation with rational numbers is limited to the last six levels of the scale and are more difficult than mental calculation with natural numbers.

Callingham and Watson (2004, pp. 80–81), highlight that the tasks suggested in these studies were given in the context of a timed mental calculation test, not within situations in which students were able to write, sketch or use an algorithm.

Their first general observation is that tasks with fractions are easier than tasks with decimals and percentages, as shown by the classification of similar tasks with different representations. One of the first acquisitions is fractions, and, in particular, the fraction of the half which appears at level A. This is probably an effect of the school curriculum,[1] which introduces fractions before decimals and percentages, or it may reflect the common use of language for half and quarter, which helps students develop simple calculations with fractions. As we advance to higher levels in fractions, denominators become bigger, non-unit fractions appear, as well as additions and subtractions with non-equivalent fractions. In level C, a unit fraction of a number appears, and division with fractions does not appear until level D. Multiplication with non-unit fractions appears at level E and operations with non-equivalent fractions at level F.

Callingham and Watson (2004, p. 80) point out that different numerical representations create additional difficulty for mental calculation with decimals, as shown by the fact that tasks with decimals appear at higher levels than equivalent tasks with fractions. The authors also highlight that in this study, all the questions with percentages requested the simple and direct calculation of X% out of Y, where Y is a whole number. The difficulty of the exercises increases as we move from percentages equivalent to fractions of half and of quarter towards multipliers of 10% and the use of integers, where the values of fractions equivalent to the percentages are not directly multipliers or divisors (e.g., 150% of 24, 90% of 40 or 30% of 80).

Another conclusion drawn by Callingham and Watson (2004, p. 81) is that the operations of multiplication and division with rational numbers are in general more difficult for students than addition and subtraction. Multiplying a fraction in its simplest form (e.g., 3 × 2/3) is not displayed until level E: it appears after multiplying a simple decimal by a power of 10 (e.g., 0.01 × 100), which occurs at level D.

Yang (2005) assesses the number sense of sixth graders in Taiwan with a series of questions, related to whole and decimal numbers. Results indicated that, regardless of performance level, very few number-sense strategies (e.g., using benchmarks, estimation or numbers of magnitude) were used. The evidence also revealed that Taiwanese students tended to apply rule-based methods and standard written algorithms to explain their reasoning.

Yang et al. (2009, p. 383) have also identified teachers' lack of number sense as a cause of student deficiencies: 'Thus, children's lack of number sense may be partly due to their teachers' lack of number sense as well as not knowing how to help students develop number sense'. Yang and colleagues have also argued that pre-service teachers in Taiwan exhibit poor number sense, relying heavily on standard written algorithms (Yang, 2007; Yang et al., 2009). Although pre-service teachers are more capable of answering number-sense test items correctly than their middle-school counterparts, the majority of their answers were also obtained by written computations based on standard algorithms.

Lemonidis, Kermeli and Palaigeorgiou (2014) proposed a teaching intervention in Greece aiming at advancing students' understanding of mental calculations and enriching their repertoire of conceptual strategy. For this purpose, three sixth-grade teachers were first trained and, later, asked to teach operations of mental calculation with rational numbers for a period of three months.

First, the three teachers were interviewed to explore their prior experience with the subject matter and pedagogical content knowledge for mental calculation in rational numbers.

The initial interview with the teachers revealed that they were dealing with serious shortcomings in the knowledge of the content of mental calculation with rational numbers, which Ball and his colleagues (2008) called *common content knowledge*. It was also obvious that they were not aware of the knowledge, attitudes or misconceptions of students in the specific subject matter, which Ball and colleagues (2008) called *specialised content knowledge*. The lack of knowledge of the mathematical content of mental calculations with rational numbers has also been noted in other studies (Post, Harel, Behr & Lesh, 1988; Cramer & Lesh 1988; Khoury & Zazkis, 1994). Hence, it is necessary for teacher education programmes to enhance *common content knowledge* and *specialised content knowledge* of mental calculations with rational numbers.

5.3.1. Mental comparison of fractions

Post, Wachsmuth, Lesh, and Behr (1985) note that children's understanding of the order of natural numbers often negatively affects their initial understanding of the order of fractions. For some children, these misconceptions persist even after a relatively intensive instruction, based on the use of manipulative devices (p. 33).

Sowder (1988) notes that there is a clear connection between the comparison of fractions and the development of number sense. The comparison of fractions is necessary to obtain an intuitive sense of their magnitude. If a fractional number is recognised, for example, as being close to 1/3 or 1/2, then there will be a more accurate sense of its magnitude. This fractional number sense is particularly important when calculating with fractions (p. 189).

Markovits and Sowder (1994) found that about 42% of seventh graders tried to use written methods (finding a common denominator or changing the fractions to decimals) when comparing 5/7 and 5/9. About 25% of them didn't know how to solve it or gave an incorrect answer. Reys and Yang (1998) and Yang and Reys (2002) also found that there was a high percentage of sixth and eighth graders that could not meaningfully compare fractions. This is due to a lack of number sense. These results are not surprising since, as mentioned above, mental calculation with rational numbers is not included in curricula or textbooks, which give emphasis to written algorithms. This fact is reflected in the opinions and attitudes of teachers.

The results presented below derive from the research of Clarke and Roche (2009, pp. 132–135) with 323 sixth-grade students in Victoria, Australia. Students were interviewed individually on eight pairs of fractions. Unsurprisingly, the easiest pair of fractions was 3/8 and 7/8 (77.1% success) and the hardest was 3/4 and 7/9 (10.8% success). However, the success rate of the easy pair (3/8 and 7/8) is not high if we consider that the students are at the end of sixth grade and fractions had been introduced many years previously.

As stated by the authors of this study, the most difficult pair (3/4, 7/9) was difficult for many of the teachers in primary and secondary education encountered during the research. Many teachers were unable to give an explanation without converting to common denominators. Of the students who succeeded in fractional comparison, 54.3% used common denominators, while overall 40% of students used some kind of residual thinking strategy (2/8 > 2/9 or 1/4 > 2/9) and 5.7% of students converted the fractions to decimals. Incorrect responses, at 35.6%, (among which 3/4 is chosen as bigger) illustrate gap thinking: there is a smaller gap between 3 and 4, so 3/4 is bigger.

Benchmarks and residual thinking strategies are used by students who have a deeper conceptual understanding of the magnitudes of fractions. It also seems that these strategies are not used much by the teachers who participated in the survey. These strategies emerged at high levels; they are suitable for the pair of 3/7 and 5/8 and the pair 5/6 and 7/8 as well. In addition, in the pairs above, 28.8% and 45.8% of students, respectively, used the strategy of common denominators.

According to the authors, the comparison of the pairs 4/7 and 4/5 was much more difficult (a success rate of only 37.2%) than predicted. As Mamede, Nunes and Bryant (2005) explain, in contrast with the pairs 3/8 and 7/8, in pairs 4/7 and 4/5 students think about the inverse relationship between the denominator and the quantity the fraction represents (p. 282). Among students who chose 4/7 as bigger, 73.5% of them justified their answer by saying that 'the numbers are greater', which demonstrates thinking that uses the logic of integers.

5.3.2. Instructional interventions in number sense

Number sense can be improved through instruction. Relevant studies with both students (Markovits & Sowder, 1994; Yang, 2002, 2003; Yang, Hsu & Huang, 2004) and teachers (Kaminski, 2002; Whitacre, 2012, 2014; Whitacre &

Nickerson, 2006) have demonstrated that number sense can be improved through instruction. For example, Yang (2003) reported a semester-long (about four and a half months) quasi-experimental study of two fifth-grade classes in Taiwan. Number-sense activities were conducted in the experimental class, while the control class followed the standard mathematics curriculum. The authors used the Number Sense Rating Scale (Hsu, Yang & Li, 2001) and found that the scores of the experimental class increased by 44% after instruction, while the scores of the control class increased by only 10%. Interviews showed that students from the experimental class used a higher proportion of number-sense strategies in post-instruction and retention interviews. Whitacre (2012, p. 34), in his literature review, concluded that: '(1) number sense can be improved through instruction, and (2) there is more to be learned regarding how number sense improves with instruction'.

5.4. Research results of the School of Nature and Life Mathematics on rational numbers

5.4.1. Results from the programme of teachers' distance training and experimentation in mental calculation

The research results presented below come from the programme of teachers' distance training and experimental implementation of mental calculations, introduced in Section 3.5.1. We will present the results from this research on mental calculation on fractions, decimals and percentages.

As we have already mentioned, in this study there was an initial measurement of student performance in mental calculations and then a second measurement for the same questions, after a teaching intervention performed by classroom teachers. We will present data from the following grades: the fourth (N = 138) as well as the fifth and sixth (N = 308).

5.4.1.1. Presentation of data from the initial and final questionnaire in fourth grade

Table 5.2, below, presents the success rates in three operations with decimals, and the percentages of strategies used before and after the teaching intervention. The percentages after the teaching intervention are presented in bold.

Accuracy

As can be seen in Table 5.2, the addition of 0.25 + 0.25 is easier than either the addition 0.5 + 0.75 or the subtraction 1.5 − 0.25 and has a double success rate. The addition 0.5 + 0.75 and the subtraction 1.5 − 0.25 are difficult operations for students; only a third of them were able to respond correctly.

The teaching intervention causes a statistically significant improvement in students' performance only for the subtraction 1.5 − 0.25 ($z = 2.8$, $p < 0.01$).

TABLE 5.2 Success rates and strategies used by students before and after the experimental teaching in grade 4

Operation	Accuracy N = 138		Rule-based strategy		Number-sense strategy		Indifferent strategy	
0.25 + 0.25	111	118	68	101	28	9	10	2
	80.4%	85.5%	49.3%	73.2%	20.3%	6.5%	7.2%	1.4%
0.5 + 0.75	49	55	32	39	5	7	11	7
	35.5%	39.9%	23.2%	28.3%	3.6%	5%	8%	5%
1.5 − 0.25	47	70	32	48	12	11	12	10
	34%	50.7%	23.2%	34.8%	8.7%	8%	8.7%	7.2%

Use of strategies

We have already mentioned in previous chapters that *number-sense strategies* are described as those revealing conceptual understanding and showing that the basic structures underlying numbers and operations have been understood. Answers showing conceptual understanding of the questions examined here are, for example, the following:

> In 0.25 + 0.25: 'It's 0.5. 0.25 is 1/4 of 1, 1/4 plus 1/4 equals 1/2, so it's 0.5'.
>
> In 0.5 + 0.75 → 1.25: 'I subtracted 0.25 from 0.75 to fill in 0.5 in order to result in 1, and I added the remainder'.
>
> In 1.5 − 0.25 → 1.25: 'Since 0.5 is 0.25 plus 0.25, if I subtract 0.25 from 1.5, there remains 1.25'.

Those strategies are described as *rule-based* wherein students use methods learned by heart and explanations that show a conceptual understanding of the procedure are not shown. For instance, such responses are the following:

> In 0.25 + 0.25: 'It's 0.50. If I add 25 plus 25 I find 50, if I place the decimal point I'll get 0.50'.
>
> In 0.5 + 0.75 → 1.25: 'I add vertically, I have 5, 2 and 1, and I place the decimal point after 1'.

An *indifferent strategy* describes a response that does not show any interest or cannot be classified with either of the two categories above. Extensively, *indifferent* are responses wherein the student's strategy leads nowhere or he confuses the method he was taught to solve the corresponding operation. For example, for 0.5 + 0.75 the answer of 0.80 is given.

According to Table 5.2, above, we observe that the students' strategies are mainly rule-based, and few students use number-sense strategies. Only in the operation 0.25 + 0.25 was there a considerable number of students (20.3%) who used number-sense strategies. In this operation, the type of numbers gave favour to their being seen as half of a half, i.e., two quarters, which are equal to a half. We also

Rational numbers **173**

observe that teaching intervention did not influence strategies used by students in favour of number-sense strategies. On the contrary, for the operation 0.25 + 0.25, after teaching intervention, the percentage of students using number-sense strategies decreased and the percentage of rule-based strategies increased.

5.4.1.2. Presentation of data from the initial and final questionnaire in grades five and six

In the following Table 5.3 we present the success rates of fifth- and sixth-grade students in operations with decimals, fractions and percentages and the percentages of strategies used before and after the teaching intervention.

Accuracy

In Table 5.3 we can observe that success rates for rational numbers are low in general: these rates range from 55% to 62%, i.e., almost half of all students were not able solve these exercises correctly. The teaching intervention improves the success rates statistically significantly in all the tasks, which then range from 61.5% to 75.5%.

Use of strategies

Responses like the following were considered to be number-sense strategies:

Conversion of fractions to decimal numbers, e.g., in $1/2 + 1/4 \rightarrow 0.5 + 0.25 = 0.75$ or $3/4$. In $1/2 \div 1/4 \rightarrow 0.5 \div 0.25 = 2$.

TABLE 5.3 Success rates and strategies used by students before and after the teaching intervention in grades 5 and 6

Operation	Accuracy N = 308		Rule-based strategy		Number-sense strategy		Indifferent strategy	
1/2 + 1/4	169 / 54.9%	189 / 61.4%	63 / 20.5%	101 / 32.8%	69 / 22.4%	70 / 22.7%	16 / 5.2%	11 / 3.6%
1 − 1/4	178 / 57.8%	200 / 65%	66 / 21.4%	100 / 32.5%	76 / 24.7%	80 / 26%	14 / 4.5%	14 / 4.5%
1/2 : 1/4	173 / 56.2%	207 / 67.2%	80 / 26%	117 / 38%	71 / 23%	61 / 19.8%	11 / 3.6%	19 / 6.2%
0.5 + 0.75	179 / 58.1%	229 / 74.4%	94 / 30.5%	109 / 35.4%	57 / 18.5%	87 / 28.2%	19 / 6.2%	20 / 6.5%
1.5 − 0.25	191 / 62%	233 / 75.6%	95 / 30.8%	121 / 39.3%	58 / 18.8%	75 / 24.3%	23 / 7.5%	22 / 7.1%
25% of 80	178 / 57.8%	222 / 72.1%	80 / 26%	123 / 40%	65 / 21.1%	72 / 23.4%	22 / 7.1%	15 / 4.9%
90% of 40	168 / 54.5%	208 / 67.5%	93 / 30.2%	118 / 38.3%	46 / 15%	54 / 17.5%	21 / 6.8%	20 / 6.5%

Use of a representation of a fraction to give meaning to it, e.g.:

> For 1/2 + 1/4: '3/4, I think of the clock cycle, the half is 2 quarters and another one gives 3 quarters'.
>
> For 1 − 1/4 → '3/4, from a circle with four pieces we withdraw one'.
>
> For 1/2 ÷ 1/4 → '2, because 1/2 is the double amount of 1/4. If I've got a pizza and split it, I take the half, then split it in the middle and I take 1/4'.
>
> For 0.5 + 0.75 → '1.25. I subtracted 0.25 from 0.5 to fill in 0.75 to get 1, and I added the remainder.
>
> For 1.5 - 0.25 → '1.25. Since 0.5 is 0.25 plus 0.25, if I subtract 0.25 from 1.5, there remains 1.25'.
>
> For 25% of 80: In the strategy of conversion to fraction, 25% can be considered as the fraction of 1/4. '25% equals 1/4, this is why it is 20'. There is also the strategy of a known factor or halving. We know 50%, half of it is 25%.

Similarly for 90% of 40. The strategy of benchmarks is used; we know that 10% of 40 gives 4, 40 − 4 = 36.

We have considered the responses such as the following as rule-based:

> For 1/2 + 1/4 → '3/4, 1/2 is 2/4 and I added 1/4'.
>
> For 1 − 1/4 → '3/4, 1 is 4/4, minus 1/4 we've got 3/4'.
>
> For 1/2 ÷ 1/4 → '2, instead of doing 1/2 ÷ 1/4, I multiplied 1/2 by 4/1'.
>
> For 0.5 + 0.75 → '1.25. I add vertically, we've got 5, 2 and 1, I place the decimal point after 1'.
>
> In 25% of 80: the algorithm is implemented for the calculation, 25/100 × 80.
>
> In 90% of 40: the algorithm taught at school is used.

Based on the data in Table 5.3, we observe that operations in which the students use rule-based strategies at higher rates are operations with decimal numbers (0.5 + 0.75 and 1.5 − 0.25), as well as the operation with the percent (90% of 40). In the operation using a percentage (25% of 80), as well as in operations with fractions, the rates of use of rule-based and number-sense strategies are almost identical. Number-sense strategies are used because the numbers 25%, 1/2 and 1/4 can be easily converted into corresponding fractional or decimal numbers (1/4, 0.5 and 0.25). Overall, our experience and the results of other investigations find that Greek students use rule-based strategies to mentally calculate rational numbers, because mental calculations with rational numbers are not instructed in school and students are taught only written rules for calculations. However, the method of converting fractions to decimals is explicitly instructed. Therefore, it may be that the strategy of converting fractions to decimals should be considered a rule-based strategy for Greek students, despite the fact that in the coding above it was registered as a number-sense strategy. This may be the reason why high rates of number-sense strategies appear in some operations.

5.4.2. Results of study of fifth- and sixth-grade Greek students

In a study conducted by Lemonidis and Kaiafa (2014a, 2014b), the authors attempted to examine the students' number sense concerning mental calculations with fractions and percentages, namely, to examine the performance and errors of students in operations, to record the strategies which the students use when performing mental calculations with fractions and percentages.

The sample consisted of 462 fifth- and sixth-grade students who participated in the sixth competition of Nature and Life Mathematics conducted in Greece. There were 290 fifth graders and 172 sixth graders. Students who took part in this competition were not selected as participation in the competition was completely voluntary. The positive attitude shown towards mathematics was perhaps their most distinctive feature. Although participation in the competition is voluntary, students' performance in mathematics is expected to be higher than mean performance.

Finally, the students were not taught mental calculation strategies for rational numbers. Thus, the number-sense strategies they used were spontaneous and self-developed.

A written examination was held, in which exercises with mental calculations were one of the four topics for the exam competition for each class. The other three questions were word problems. In each exercise with mental calculations, students were asked to provide two ways of solving it and, in addition, to describe their thinking in written form.

5.4.2.1. The questions

The questions concerning mental calculations that were posed to the students are the following:

Fifth grade

> **Q51:** I mentally calculate how much $1 - 1/4$ is. I answer in two ways. I write down the way I thought for both.
>
> **Q52:** I mentally calculate how much $1/2 \div 1/4$ is. I answer in two ways. I write down the way I thought for both.

Sixth grade

> **Q61:** I compare the fractions $3/7$ and $5/8$. Which is bigger? I answer in two ways. I write down the way I thought for both.
>
> **Q62:** I find 90% of 40. I answer in two ways. I write down the way I thought for both.

5.4.2.2. Computational strategies used by students

One result drawn from this study is that students use rule-based strategies in general; they used algorithms or typical rules for operations of calculation.

In Table 5.1 above, the strategies used by students in order to answer the questions asked are shown.

In Table 5.4, the percentages of students who selected each strategy are shown. Here, we observe that the majority of students perform rule-based strategies to find the result, despite the fact that they were explicitly asked to calculate *mentally*. Specifically, the percentage of the students who chose rule-based strategies as a first or second option reaches 67% for the subtraction $1 - 1/4$ among fifth graders and 80.5% for the division $1/2 \div 1/4$ among fifth graders, 58% in comparing fractions (3/7 and 5/8) among sixth graders and 62.5% in finding the percentage 90% of 40 among sixth graders.

This behaviour, namely the widespread use of rule-based strategies by students, may be justified by the 'didactic contract' (Brousseau, 1984), which is formed in classrooms where students are encouraged to use only one method (rule of operation).

The percentage of students who chose to convert from fraction or percent to decimal ranges from 16.5% to 30.5%. Number-sense strategies, namely using mental representations, using 1/2 as a reference point and residual thinking, seem to be used significantly less often.

5.4.2.3. Student mistakes

Most errors are due either to the inability of a student to correctly perform a written algorithm that has been taught or to the fact that the written algorithm of another operation is attempted. In subtraction of fractions, for example, many

TABLE 5.4 Percentages of strategies used by students answering one or two questions correctly (from Lemonidis & Kaiafa, 2014a)

Strategy	Q51: $1 - 1/4$ $N = 269$	Q52: $1/2 \div 1/4$ $N = 230$	Q61: 3/7 and 5/8 $N = 210$	Q62: 90% of 40 $N = 224$
Rule-based strategy	180 (67%)	185 (80.5%)	122 (58%)	140 (62.5%)
Number-sense strategy	89 (33%)	45 (19.5%)	88 (42%)	84 (37.5%)
Number-sense strategies				
Converting a fraction or a percent to a decimal	82 (30.5%)	38 (16.5%)	56 (26.5%)	42 (18.75%)
Mental picture	7 (2.5%)		18 (8,5%)	
Benchmarks to 1/2 or 10%		7 (3%)	7 (3.5%)	19 (8.5%)
Residual thinking			7 (3.5%)	
Reduction in unit				23 (10.3%)

students reversed the terms of the second fraction and then performed multiplication (influenced by the algorithm of dividing fractions), while others, in division, converted to fractions with the same denominators, divided the numerators and preserved the same denominator, clearly influenced by the written algorithm of addition and subtraction of fractions. Finally, in the comparison of fractions, some students focused on the numerators or denominators only in order to compare, e.g., 'Between fractions 3/7 and 5/8, 5/8 is bigger, because its numerator is larger', or '3/7 is bigger, because its denominator is smaller'.

The authors of this study conclude that student mistakes are attributable to not having developed an intuitive knowledge and understanding of rational numbers and make mistakes while trying to perform a written algorithm whose meaning and function cannot be understood.

5.4.3. Instructional intervention for sixth-grade students

Lemonidis, Kermeli and Palaigeorgiou (2014) proposed a remedial instruction in mental calculation with rational numbers of sixth-grade Greek students. One of the main questions that this study pursued was: How did the teaching intervention alter the strategies for rational numbers that were used by the sixth-grade students? Questionnaires containing four problems of mental calculation with rational numbers were given to the 66 students. The problems were identical to the ones given to 462 students of the fifth and sixth grades who participated in the mathematics contest. These students had not been taught mental calculation strategies with rational numbers and, thus, the number-sense strategies they used were spontaneous and self-developed. The four questions that students of both groups had to answer, were:

I mentally calculate: Q1: $1 - 1/4$, Q2: $1/2 \div 1/4$, Q3: I compare the fractions 3/7 and 5/8, Q4: I find 90% of 40. I use two ways to answer. Each time, I write the way I thought.

TABLE 5.5 Strategy use of students who succeeded in one or two questions

Strategy	Q1: $1 - 1/4$ $N = 63$ $N = 269$	Q2: $1/2:1/4$ $N = 59$ $N = 230$	Q3: 3/7 and 5/8 $N = 48$ $N = 210$	Q4: 90% of 40 $N = 61$ $N = 224$
Rule-based strategy				
Experimental group	25 (39.7%)	31 (52.5%)	17 (35.4%)	33 (54%)
Control group	180 (67%)	185 (80.5%)	122 (58%)	140 (62.5%)
Number-sense strategy				
Experimental group	38 (60.3%)	28 (47.5%)	31 (64.6%)	28 (45.9%)
Control group	89 (33%)	45 (19.5%)	88 (41.9%)	84 (37.5%)
	z = 4 $p < 0.001$	z = 4.39 $p < 0.001$	z = 2.84 $p < 0.01$	z = 1.19 $p > 0.05$

5.4.3.1. Students' use of strategies after interventional instruction

The following table shows the percentages of rule-based and number-sense strategies used by the students (experimental group) who participated in the study and students (control group) who participated in the competition.

The students in the experimental group used a smaller percentage of rule-based strategies and a larger percentage of number-sense strategies in comparison to the students of the control group. A 2-sample z-test validated this observation by indicating that in three (Q1, Q2 and Q3) of the four questions, the difference was significant. The proposed instruction promoted number-sense strategies in operations on rational numbers. Students from the experimental group understood the operations with rational numbers better, despite the fact that students who participated in the contest may have had a more positive attitude towards mathematics.

Another effect was that the respective students used a greater variety of number-sense strategies.

Note

1 This refers to the Australian curriculum.

References

Ball, D. L., Thames, M. H. & Phelps, G. (2008). Content knowledge for teaching: What makes it special? *Journal of Teacher Education*, 59, 389–407.

Behr, M. J., Wachsmuth, I., Post, T. R. & Lesh, R. (1984). Order and equivalence of rational numbers: A clinical teaching experiment. *Journal for Research in Mathematics Education*, 15, 323–341.

Booth, L. R. (1981). Child-methods in secondary mathematics. *Educational Studies in Mathematics*, 12, 29–41.

Bright, G. W., Behr, M. J., Post, T. R. & Wachsmuth, I. (1988). Identifying fractions on number lines. *Journal for Research in Mathematics Education*, 19(3), 215–232.

Brousseau, G. (1984). The crucial role of the didactical contract in the analysis and construction of situations in teaching and learning mathematics. In H.-G. Steiner (Ed.), *Theory of mathematics education: ICME 5 – topic area and miniconference: Adelaide, Australia*. Bielefeld, Germany: Institut fuer Didaktik der Mathematik der Universitaet Bielefeld.

Brousseau, G., Brousseau, N. & Warfield, V. (2007). Rationals and decimals as required in the school curriculum: Part 2: From rationales to decimals. *Journal of Mathematical Behavior*, 26, 281–300.

Callingham, R. & Watson, J. (2004). A developmental scale of mental computation with part-whole numbers. *Mathematics Education Research Journal*, 16(2), 69–86.

Callingham, R. & Watson, J. (2008). *Research in mental computation: Multiple perspectives*. Brisbane: Post Pressed. ISBN: 978-1-921214-36-3.

Caney, A. & Watson, J. M. (2003). *Mental computation for part-whole number operations*. Paper presented at the joint conferences of the Australian Association for Research in Education and the New Zealand Association for Research in Education, Auckland. Retrieved from http://www.aare.edu.au/03pap/alpha.htm September 3, 2004.

Clarke, D. M. & Roche A. (2009). Students' fraction comparison strategies as a window into robust understanding and possible pointers for instruction. *Educational Studies in Mathematics*, 72(1), 127–138.

Cobb, P. (1999). Individual and collective mathematical development: The case of statistical data analysis. *Mathematical Thinking and Learning*, 1, 5–43.

Cramer, K. & Lesh, R. (1988). Rational number knowledge of preservice elementary education teachers. In M. Behr (Ed.), *Proceedings of the 10th Annual Meeting of the North American Chapter of the International Group for Psychology of Mathematics Education* (pp. 425–431). DeKalb, Il.: PME.

Dole, S. (1999). Successful percent problem solving for year 8 students using the proportional number line method. In J. M. Truran & K. M. Truran (Eds.), *Making the difference (Proceedings of the 22nd annual conference of the Mathematics Education Research Group of Australasia*, Vol. 1, pp. 43–50). Sydney: MERGA.

Dufour-Janvier, B., Bednarz, N., & Belanger, M. (1987). Pedagogical considerations concerning the problem of representation. In C. Janvier (Ed.), *Problems of representation in the teaching and learning mathematics* (pp. 109–122). Hillsdale, NJ: Lawrence Erlbaum Associates, Inc.

Geary, D. C. (2006). Development of mathematical understanding. In W. Damon, R. Lerner, D. Kuhn & R. Siegler (Eds.), *Handbook of child psychology: Cognition, perception, and language*, Vol. 2 (pp. 777–810). Hoboken, NH: Wiley.

Gelman, R. & Williams, E. (1998). Enabling constraints for cognitive development and learning: Domain specificity and epigenesis. In W. Damon, D. Kuhn & R. S. Siegler (Eds.), *Handbook of child psychology*. New York: Wiley.

Harel, G. & Confrey, J. (Eds.). (1994). *The development of multiplicative reasoning in the learning of mathematics*. Albany, NY: SUNY Press.

Hart, K., Brown, M., Kerslake, D., Kuchermann, D. & Ruddock, G. (1985). *Chelsea Diagnostic Mathematics Tests. Fractions 1*. Windsor, UK: NFER-Nelson.

Hsu, C. Y., Yang, D. C. & Li, F. M. (2001). The design of the fifth and sixth grade number sense rating scale. *Chinese Journal of Science Education (TW)*, 9, 351–374.

Kaminski, E. (2002). Promoting mathematical understanding: Number sense in action. *Mathematics Education Research Journal*, 14, 133–149.

Kerslake, D. (1986). *Fractions: Children's strategies and errors: A report of the strategies and error in secondary mathematics project*. Windsor, UK: NFER-Nelson.

Kieren, T. E. (1993). Rational and fractional numbers: From quotient fields to recursive understanding. In T. P. Carpenter, E. Fennema & T. A. Romberg (Eds.), *Rational numbers: An integration of research* (pp. 49–84). Hillsdale, NJ: Lawrence Erlbaum Associates, Inc.

Khoury, H. A. & Zazkis, R. (1994). On fractions and non-standard representations: Preservice teachers' concepts. *Educational Studies in Mathematics*, 27, 191–204.

Lamon, S. J. (1999). *Teaching fractions and ratios for understanding: Essential content knowledge and instructional strategies for teachers*. Mahwah, NJ: Lawrence Erlbaum.

Lembke, L. O. & Reys, B. J. (1994). The development of, and interaction between, intuitive and school-taught ideas about percent. *Journal for Research in Mathematics Education*, 25, 237–259.

Lemonidis, Ch. & Kaiafa, I. (2014a). Fifth and sixth grade students' number sense in rational numbers and its relation with problem solving ability. *MENON: Journal Of Educational Research. 1st Thematic Issue*, 61–74.

Lemonidis, Ch. & Kaiafa, I. (2014b). Κατανόηση και ευελιξία των μαθητών Ε' και Στ' τάξης στους υπολογισμούς με ρητούς αριθμούς. [Understanding and flexibility of students fifth and sixth grades in the calculations with rational numbers.] Πρακτικά 5ου

Συνεδρίου Ένωσης Ερευνητών Διδακτικής των Μαθηματικών (Ε.ΝΕ.ΔΙ.Μ) ΕΝΕΔΙΜ, Φλώρινα. *Proceedings of the 5th Congress Union Researchers Teaching of Mathematics (E.NE. DI.M) 5th ENEDIM*, Florina, 2014.

Lemonidis, Ch., Kermeli, A. & Palaigeorgiou, G. (2014). Exploring number sense in sixth grade in Greece: An instructional proposal and its learning results. *MENON: Journal Of Educational Research, 1st Thematic Issue*, 159–172.

Lesh, R., Behr, M. & Post, T. R. (1987). Rational number relations and proportions. In C. Janvier (Ed.). *Problems of representation in the teaching and learning mathematics* (pp. 41–58). Hillsdale, NJ: Lawrence Erlbaum Associates, Inc.

Leslie, A. M., Gelman, R. & Gallistel, C. R. (2008). The generative basis of natural number concepts. *Trends in Cognitive Sciences, 12*(6), 213–218.

Madison, B. L. & Steen, L. A. (Eds.) (2003). *Quantitative literacy: Why numeracy matters for schools and colleges*. Princeton, NJ: The National Council on Education and the Disciplines.

Mack, N. (1995). Confounding whole-number and fraction concepts when building on informal knowledge. *Journal for Research in Mathematics Education, 26*, 422–441.

Mamede, E., Nunes, T. & Bryant, P. (2005). The equivalence of ordering of fractions in part-whole and quotient situations. In H. L. Chick & J. L. Vincent (Eds.), *Proceedings of the 29th Conference of the International Group for the Psychology of Mathematics Education* (pp. 281–288). Melbourne: PME.

McIntosh, A. J., De Nardi, E. & Swan, P. (1994). *Think mathematically*. Melbourne: Longman.

Markovits, Z. & Sowder, J. (1991). Students' understanding of the relationship between fractions and decimals. *Focus on Learning Problems in Mathematics, 13*(1), 3–11.

Markovits, Z. & Sowder, J. T. (1994). Developing number sense: An intervention study in grade 7. *Journal for Research in Mathematics Education, 25*(1), 4–29.

Moskal, B. M. & Magone, M. E. (2000). Making sense of what students know: Examining the referents, relationships and modes students displayed in response to a decimal task. *Educational Studies in Mathematics, 43*, 313–335.

Moss, J. (2005). Pipes, tubes, and beakers: New approaches to teaching the rational-number system. In M. S. Donovan & J. D. Bransford (Eds.), *How students learn: Mathematics in the classroom* (pp. 121–162). Washington, DC: National Academic Press.

National Council of Teachers of Mathematics (NCTM) (2000). *Principles and standards for school mathematics*. Reston, VA: Author.

Ni, Y. & Zhou, Y.-D. (2005). Teaching and learning fraction and rational numbers: The origins and implications of whole number bias. *Educational Psychologist, 40*(1), 27–52.

Noelting, G. (1980a). The development of proportional reasoning and the ratio concept. Part 2 – Problem structure at successive stages: Problem-solving strategies and the mechanism of adaptive restructuring. *Educational Studies in Mathematics, 11*(3), 331–363.

Noelting, G. (1980b). The development of proportional reasoning and the ratio concept. Part 1 – differentiation of stages. *Educational Studies in Mathematics, 11*(2), 217–253.

Nunes, T. & Bryant, P. (2009). Paper 3: Understanding rational numbers and intensive quantities. In T. Nunes, P. Bryant & A. Watson (Eds.), *Key understandings in mathematics learning*. London: Nuffield Foundation.

Nunes, T., Bryant, P., Pretzlik, U. & Hurry, J. (2006). *Fractions: Difficult but crucial in mathematics learning*. London Institute of Education, London: ESRC-Teaching and Learning Research Programme.

O'Connor, M. C. (2001). 'Can any fraction be turned into a decimal?' A case study of a mathematical group discussion. *Educational Studies in Mathematics, 46*, 143–185.

Post, T., Behr, M. J. & Lesh, R. (1986). Research-based observations about children's learning of rational number concepts. *Focus on Learning Problems in Mathematics*, 8(1), 39–48.

Post, T., Cramer, K., Behr, M., Lesh, R. & Harel, G. (1993). Curriculum implications of research on the learning, teaching and assessing of rational number concepts. In T. Carpenter, E. Fennema & T. Romberg (Eds.), *Rational numbers: An integration of research* (pp. 327–361). Hillsdale, NJ: Lawrence Erlbaum.

Post, T., Harel, G., Behr, M. & Lesh, R. (1988). Intermediate teachers knowledge of rational number concepts. In E. Fennema, T. Carpenter & S. Lamon (Eds.), *Papers from the first Wisconsin symposium for research on teaching and learning mathematics* (pp. 194–219). Madison, WI: Wisconsin Center for Education Research.

Post, T., Wachsmuth, I., Lesh, R. & Behr, M. (1985). Order and equivalence of rational numbers: A cognitive analysis. *Journal for Research in Mathematics Education*, 16(1), 18–36.

Reys, R. E. & Yang, D. C. (1998). Relationship between computational performance and number sense among sixth- and eighth-grade students in Taiwan. *Journal for Research in Mathematics Education*, 29, 225–237.

Shaughnessy, J. M. (2003). Research on students' understandings of probability. In J. Kilpatrick, W. G. Martin & D. Schifter (Eds.), *A research companion to principles and standards for school mathematics* (pp. 216–226). Reston, VA: National Council of Teachers of Mathematics.

Skemp, R. R. (1976). Relational understanding and instrumental understanding. *Mathematics Teaching*, 77, 20–26.

Sowder, J. T. (1988). Mental computation and number comparisons: The role in development of number sense and computational estimation. In J. Hiebert & M. Behr (Eds.), *Number concepts and operations in the middle grades* (pp. 182–197). Reston, VA: Lawrence Erlbaum and National Council of Teachers of Mathematics.

Stacey, K. & Steinle, V. (1998). Refining the classifications of students' interpretations of decimal notation. *Hiroshima Journal of Mathematics Education*, 6, 49–69.

Stafylidou, S. & Vosniadou, S. (2004). The development of students' understanding of the numerical value of fractions. *Learning and Instruction*, 14, 503–518.

Steen, L. A. (Ed.) (2001). *Mathematics and democracy: The case for quantitative literacy.* Washington, DC: National Council on Education and the Disciplines.

Thompson, P. W. & Saldanha, L. A. (2003). Fractions and multiplicative reasoning. In J. Kilpatrick, W. G. Martin & D. Schifter (Eds.), *A research companion to principles and standards for school mathematics* (pp. 95–113). Reston, VA: National Council of Teachers of Mathematics.

Vamvakoussi, X. & Vosniadou, S. (2004). Understanding the structure of the set of rational numbers: A conceptual change approach. *Learning and Instruction*, 14, 453–467.

Vamakoussi, X. & Vosniadou, S. (2010). How many decimals are there between two fractions? Aspects of secondary school students' understanding of rational numbers and their notation. *Cognition and Instruction*, 28, 181–209.

Vosniadou, S., Vamvakoussi, X. & Skopeliti, I. (2008). The framework theory approach to conceptual change. In S. Vosniadou (Ed.), *International handbook of research on conceptual change* (pp. 3–34). Mahwah, NJ: Erlbaum.

Vergnaud, G. (1997). The nature of mathematical concepts. In T. Nunes and P. Bryant (Eds.), *Learning and teaching mathematics. An international perspective* (pp. 1–28). Hove (UK): Psychology Press.

Watson, J. M. (2004). Quantitative literacy in the media: An arena for problem solving. *Australian Mathematics Teacher*, 60(1), 34–40.

Watson, J. M., Collis, K. F. & Campbell, K. J. (1995). Developmental structure in the understanding of common and decimal fractions. *Focus on Learning Problems in Mathematics*, *17*(1), 1–24.

Watson, J. M., Collis, K. F. & Moritz, J. B. (1997). The development of chance measurement. *Mathematics Education Research Journal*, *9*, 60–82.

Whitacre, I. (2012). *Investigating number sense development in a mathematics content course for prospective elementary teachers* (unpublished doctoral dissertation). University of California, San Diego, and San Diego State University.

Whitacre, I. (2014). Strategy ranges: Describing change in prospective elementary teachers' approaches to mental computation of sums and differences. *Journal of Mathematics Teacher Education*, *18*(4), 353–373.

Whitacre, I. & Nickerson, S. D. (2006). Pedagogy that makes (number) sense: A classroom teaching experiment around mental math. In S. Alatorre, J. L. Cortina, M. Sáiz & A. Méndez (Eds.), *Proceedings of the twenty-eighth annual meeting of the North American Chapter of the International Group for the Psychology of Mathematics Education, Vol. 2* (pp. 736–743). Mérida, México: Universidad Pedagógica Nacional.

Wynn, K. (2002). Do infants have numerical expectations or just perceptual preferences? *Developmental Science*, *2*, 207–209.

Yang, D. C. (2002). Teaching and learning number sense: One successful process-oriented activity with sixth grade students in Taiwan. *School Science and Mathematics*, *102*(4), 152–157.

Yang, D. C. (2003). Teaching and learning number sense – An intervention study of fifth grade students in Taiwan. *International Journal of Science and Mathematics Education*, *1*(1), 115–134.

Yang, D. C. (2005). Number sense strategies used by sixth grade students in Taiwan. *Educational Studies*, *31*(3), 317–334.

Yang, D. C. (2007). Investigating the strategies used by preservice teachers in Taiwan when responding to number sense questions. *School Science and Mathematics*, *107*, 293–301.

Yang, D. C., Hsu, C. J. & Huang, M. C. (2004). A study of teaching and learning number sense for sixth grade students in Taiwan. *International Journal of Science and Mathematics Education*, *2*(3), 407–430.

Yang, D. C. & Reys, R. E. (2002). Fractional number sense strategies possessed by sixth grade students in Taiwan. *Hiroshima Journal of Mathematics Education*, *10*(2), 53–70.

Yang, D. C., Reys, R. E. & Reys, B. J. (2009). Number sense strategies used by pre-service teachers in Taiwan. *International Journal of Science and Mathematics Education*, *7*(2), 383–403.

6
COMPUTATIONAL ESTIMATION

Estimation is a process widespread in the everyday life of children and adults. We often need to do quick calculations or judgements of numerical magnitudes, without using a computer or paper and pencil, to answer questions such as: Do I have enough money to buy what I need? How much time do we still need to get home? How heavy is it? Approximately how many people were at the meeting? Estimation may be used in everyday life more often than any other quantification. Estimation is recognised as an important mathematical content and is highlighted in the curricula of many countries, such as England, the United States of America, Japan, etc. In the *Principles and standards for school mathematics* (NCTM, 2000) of the United States, the importance of skills development in computational estimation is recognised as well.

A second reason why estimation is important, apart from its usefulness in everyday life, is that it can help students develop a better understanding of place value and of operations and a better number sense in general. Computational estimation is a main element of number sense (Greeno, 1991; McIntosh, 2004).

A third reason is that many types of estimation require us to go further than the certain application of fixed procedures to the application of mathematical knowledge in flexible ways. This adaptive type of problem solving is one of the main goals of the modern teaching of mathematics.

A last reason for the practical significance of estimation is that students (e.g., Hanson & Hogan, 2000; Sigler & Booth, 2004; Tsao & Pan, 2011), pre-service (e.g., Castro, Castro & Sergovia, 2002; Lemonidis & Kaimakami, 2013; Tsao, 2013) and in-service teachers (e.g., Alajmi, 2009; Dowker, 1992; Mildenhall, Hackling & Swan, 2009; Tsao & Pan, 2013) and adults (e.g., Anestakis & Lemonidis, 2014; Lemaire, Arnaud & Lecacheur, 2004) encounter difficulties in the process of estimation.

6.1. Definitions and various kinds of estimation

The situations of estimation that we face in our daily lives constitute different types of estimation: *computational estimation, measurement estimation, numerosity estimation* and *number line estimation* (Sowder, 1992; Siegler & Booth, 2005). Each of these types of estimation requires a different understanding and skill set.

According to Sowder (1992, p. 371),

> The most common type of computational estimation problem requires estimating the result of a computation by performing some mental computation on approximations of the original numbers. To be correct, the answer must fall within a certain interval, as determined by the problem itself or some outside source, such as a teacher.

It is also thought that computational estimation is a component of number sense (Greeno, 1991; McIntosh, 2004). The ability to estimate numerical quantities is a component of number sense (for more see Chapter 1, Section 1.2.3).

Measurement estimation is performed on objects with continuous rather than discrete magnitudes, based on an arbitrary metric unit (e.g., the measure of length). Such sizes are length, area, volume, time, weight, etc. For example, in order to estimate the length of the perimeter of a fence with pickets at a constant distance, someone might first determine the length between the pickets and then to multiply in order to estimate the length of the perimeter of the fence. Similarly, in order to calculate the length of one wall by concrete blocks, it suffices to estimate the length of a concrete block and multiply their number.

We can define measures of estimation as the process by which we arrive at a measurement without using measuring tools. Instead, *measurement* is the process of comparing a physical object's characteristic with a default unit of measurement. For a good estimation of measurement, one should have a good sense of the size of the unit used. For example, a strong ability to estimate measurements is possessed by people working in professions such as: builders, tilers, etc., where such measurements and estimations are constantly performed.

The estimation of *numerosity* is usually raised by the question 'how many?' and asked in order to find the number of elements in a set. For example, we try to estimate the number of people in a cinema hall, the number of cards in a stack of cards, how many candies there are in a box, etc. A common procedure used in estimating numerosity is to measure a sample and then multiply by the number of samples. Thus, in order to estimate the number of people inside a stadium, we can measure or estimate the number of people in a small area, to estimate the number of areas and multiply to find the whole. There are several variations of this method, such as sectioning to find the number of candies in a transparent box or working by ratios (how many candies in the box are red?) or various other methods.

The problems of estimating numerosity can become quite complex. A well-known problem in the literature of problem solution is Schoenfeld's, which seeks to estimate the number of cells in a medium-sized adult human body (Schoenfeld, 1985). Such problems are usually encountered in estimation, since the exact answer is not possible.

Number line estimation requires either the interpretation of a number in a place in space, on a number line, or the interpretation of a point on a number line with a number. The students' interpretations of numbers on positions of number lines give very direct information regarding the representations they have of numerical terms. Number line estimation steadily improves over the school year in primary education: the accuracy is greater at any given age at the lowest numerical scales. Siegler and Booth (2004) find that the students' mistakes are reduced from 24% to 10% between kindergarten and the second grade of primary school.

Siegler and Booth (2005, p. 198) give the following definition for estimation: *Estimation is a process of translating between alternative representations, at least one of which is inexact.* Quantitative representations can be either numerical or non-numerical. According to this definition, types of estimations are grouped into two major categories: numerical estimation and non-numerical estimation.

Numerical estimation corresponds to estimation tasks in which one or both sides of the translation involve numbers. For example, computational estimation requires the translation of one numerical representation (e.g., 4,893 + 5,124) into another (about 10). Estimating measures or estimating numerosity may require the translation of a non-numerical representation (e.g., the visual representation of a wall's length or the volume and density of a candy box) to a number. Number line estimation requires translation of a number to a place in space or the translation of a position in space to a number.

Non-numerical estimation involves tasks in which none of the quantitative representation is numerical. In this category are included situations found in psychophysical experiments requiring translation between two non-numerical quantitative representations, as for example between the brightness of a lamp and spatial position in a park.

6.1.1. Estimation and approximation

Many times we do not distinguish the terms estimation and approximation, so these terms coincide or are used interchangeably. Approximation usually occurs when it is impossible or difficult to obtain an exact form or an exact number. For example, the number π, approximated to four decimal places, is 3.1416, as $\sqrt{2}$ is likewise approximated to 1.4142. The approximation attempts to approach a target value.

Approximation plays an important role in measurements. Each measurement of a continuous quantity is actually an approximation, because there is not a precise value. Thus, the activities of estimation and measurement, which are often used in classrooms, require students to create an estimation and then find an approximation using measurement.

6.1.2. Computational estimation

The importance and the frequent appearance of estimation and its forms in the everyday life of children is widely known. The estimation of the time spent in a lecture, the sum of 0.765 plus 0.421 and the number of fans in a stadium have very little in common, apart from the fact that the answer for each is an estimation. This diversity of situations involving estimation may be bound up with many difficulties and many different developmental models. Thus, situations of estimation are classified based on the knowledge required for a proper estimation. The two kinds of knowledge that seem to be useful to separate the situations of estimation are the *content knowledge of real life* and *numbers*. Some situations require the knowledge of specific entities in real life, conventional measurement units, or both. One example that requires both is the estimation of kilometres between two cities. When we make estimations in additions or in multiplications, we input and output only numbers. There are situations that do not require the knowledge of reality and numbers (e.g., to estimate whether there is enough time to cross the road before the car down the street reaches us).

Thus we will focus only on computational estimations, because they have been studied more and more research data exists for this type of estimation in literature.

When we are called on to calculate in real life or in school, the first decision we must take is to determine *if an exact calculation or a computational estimation is required*. We present the reasons for which the use of estimation or computational estimation is necessary:

First, there are many situations where exactness is not necessary (Segovia & Castro, 2009, p. 506; Van den Heuvel-Panhuizen, 2001, p. 173). For example, an exact calculation is not necessary to decide if 20 Euros are enough for the purchase of three tickets of €5.95.

Second, often it is impossible to know an exact value because of the context of the situation, e.g., the exact determination of a date in antiquity; or the kind of numbers that are involved in the situation, e.g., calculating with periodic decimal numbers (Segovia & Castro, 2009, pp. 505–506; Van den Heuvel-Panhuizen, 2001, p. 173).

Third, computational estimation may serve numerical clarity, e.g., in the media Greece will be indicated as a country of 11 million inhabitants and not 10,816,286 (Segovia & Castro, 2009, p. 506).

Last, computational estimation offers ease in computation, e.g., a mental, approximate calculation is enough to produce an accurate response that will lead to making a decision (Segovia & Castro, 2009, p. 506).

Siegler and Booth (2005, p. 199) describe more concisely the various types of conceptual understanding required in an efficient computational estimation: (a) that the goal of estimation is to produce an answer reasonably close in magnitude to the correct one, (b) that approximate numbers are useful for attaining this goal, (c) that the estimation may involve multiple valid approaches and multiple reasonable answers, and (d) that the context determines the adequacy of answers.

6.2. Strategies used in computational estimation

In this section, based on various studies, we will present strategies used in computational estimation. Computational estimation strategies have been investigated in adults with different levels of computing skills (Dowker, 1992; Dowker, Flood, Griffiths, Harriss & Hook, 1996), in children and adolescents (Levine, 1982; Baroody, 1989; Sowder & Wheeler, 1989; Reys, Reys & Penafiel, 1991a; Dowker, 1997; Lemaire, Lecacheur & Farioli, 2000) and in samples of various ages (LeFevre, Greenham & Waheed, 1993; Lemaire & Lecacheur, 2002). In most of these studies, the computational estimation ability was investigated by asking participants to give estimates or approximate solutions to arithmetical problems (e.g., 246 + 468 = 700). The investigations were carried out with precision measurement approaches (measuring the absolute or the relative difference between the estimated and the correct answers), with oral protocols (e.g., by asking examinees to say how the solution was found) and in some cases by measuring response time.

The results of these studies show that in the development of computational estimation, children use a variety of strategies from a very early stage. These strategies are not often used in a unique way, but can be combined: for example, in a problem, one student may first round one or two factors and then use compensation to reduce a gap created by the rounding.

The strategies in computational estimation can be classified according to different levels of generalisation. At a general level, it has been found that children and adults use the following three groups of strategies: *reformulation, compensation* and *translation* (Reys, Bestgen, Rybolt & Wyatt, 1982; Reys, Reys, Nohda & Ishida, 1991b; Sowder & Wheeler, 1989).

Reformulation is effected in a problem where original numbers are replaced with new ones more convenient for calculation before computing. An example is reformulation in rounding, e.g., $31{,}151 + 62{,}198 \approx 31{,}000 + 62{,}000 \approx 93{,}000$.

Sowder and Wheeler (1989, p. 132) include four strategies in the group of reformulation strategies: *(a) rounding, (b) truncation, (c) averaging and (d) changing the form of a number.*

Compensation occurs when the result of an initial estimation is corrected. In the above example, initially one calculates $31{,}151 + 62{,}198$ to be 93,000 and then returns, adds 300 and finds 93,300.

Translation is a change in the structure of the problem to a more convenient form of computation. For example, to calculate a sum of a long list of numbers, we calculate the average of numbers and multiply by the number of the addends.

Then we will present a list of various strategies shown in computational estimation based on the literature (Reys et al., 1982; Sowder & Wheeler, 1989; Reys et al., 1991b; Dowker, 1992; LeFevre et al., 1993; Mildenhall, 2009). These strategies often depend on the operations given in the estimation. In some cases, the names given to them by various authors vary.

1. *Intuition:* this strategy is mainly used by younger and inexperienced estimators and can reinforce the belief that approximate answers can be valuable. For example, a student may respond that the average of 3, 5, 8 and 10 is about 6, but cannot justify his answer.
2. *Rounding:* converts one or both numbers to the nearest number that results in one or more zeros (e.g., 498 + 496, both addends can be converted to 500). May follow with an adjustment of the estimation, if greater precision is required.

   We can distinguish two types of rounding:

   - Rule-based rounding: rounding here is based on a well-known rule. We determine what the rounding digit is and check the digit on the right of it. If it's equal or greater than 5, the rounding digit is increased by one. In contrast, if it is less than 5, then the rounding digit remains the same.
   - Situation-based rounding: takes into account the context of the arithmetic problem and a specific situation of calculation. Note that it is a neither necessary nor desirable attachment to the rule of rounding. Instead, the rounding process must be flexible and be done in many different ways, to facilitate mental calculation (Reys, 1984; Reys, 1986). For example, for the product 65 × 23, the roundings 60 × 20, 70 × 20 or 70 × 23 are all acceptable.

3. *Front-end strategy:* although it may be useful in all four operations, its main usefulness is in addition. It can be done in two steps. In the first step, it focuses on the digits to the left end of numbers, ignoring the rest. After formulating the estimation, an arrangement with the calculation of segments that have been ignored may be created. For example, 50.48 + 35.13 + 705.5 + 90.89 is computed by the sum of the integer parts (front end) 50 + 35 + 705 + 90 = 880, then an arrangement can be made by summing decimal parts 0.48 + 0.13 + 0.5 + 0.89 ≈ 2 and one of the other two numbers, so in total we have 880 + 2 ≈ 882.
4. *Truncating:* this strategy has the same logic as the previous strategy, based on the relative size of the numbers and the place value of the front digit. In the truncated number the front digit or digits (those with the greater place value) stay the same, while the remaining numbers (to the right) are converted to zeros. For example, 5,682 may be converted to 5,000, 5,600 or 5,680. The selection of the position of the truncation is selected depending on the situation of calculation.
5. *Clustering or averaging:* this special strategy is applied in the addition of multiple numbers, when these numbers are near to a specific price. For example, 23 + 18 + 19 + 22 can be calculated as 4 × 20 = 80.
6. *Compensation:* we can distinguish two types of compensation, prior and post compensation. *Prior compensation:* the second term is rounded in the opposite direction from the first, before any operation. For example, in 57 × 56, 56 may become 50 instead of 60, to compensate the rounding of 57 to 60, so 60 × 50 = 3,000, while 60 × 60 = 3,600, which is further from the exact result 3,192.

Post compensation: Post compensation is also encountered in the literature as an *adjusting strategy* (Reys, 1984) and is a process that is applied after the use of a computational estimation strategy to correct the initial estimation, when greater accuracy is desired. For example, $57 \times 56 \approx 60 \times 60 = 3{,}600$; we subtract $3 \times 60 = 180$ and $4 \times 60 = 240$, removing about 400 of 3,600, obtaining 3,200.

7. *Compatible numbers strategy:* this strategy involves the selection of numbers that make an estimation easier and give a good estimate of the original problem. For example, in the sum $35 + 46 + 65 + 71 + 60 + 38$, $35 + 71$ is about 100, $46 + 60$ is approximately 100, and $65 + 38$ is approximately 100. Thus, the sum is approximately 300. These pairs of numbers are then compatible.
8. *Special numbers strategy or benchmarking:* in many cases, students are trained to distinguish numbers which are close to special values. This happens with fractions, where specific values are 0, $\frac{1}{2}$ and 1. For example, in the sum $\frac{3}{4} + \frac{1}{25} + \frac{7}{13}$, $\frac{3}{4}$ can be regarded as 1, $\frac{1}{25}$ can be regarded as 0 and $\frac{7}{13}$ can be regarded as $\frac{1}{2}$. Thus the total is approximately 1.5.
9. *Reformulation or substitution:* changing the form of one or two numbers in order to generate an easier calculation (e.g., the operation 0.52×0.35 may be converted to $\frac{1}{2} \times \frac{1}{3}$).
10. *Range:* students use this strategy to calculate the range into which the answer to a calculation is expected to fall. For example, an answer of 2.6×7 is expected to be between 14 and 21, namely the lower limit of $2 \times 7 = 14$ and the upper limit of $3 \times 7 = 21$.
11. *Factorisation:* analysing numbers into a simpler form. For example, 128×152 is $130 \times 10 \times 15$.
12. *Distributivity:* using distributivity. For example, for 38×91 we may obtain $(38 \times 100) - (38 \times 10) = 3{,}800 - 380$: approximately 3,400.
13. *Proceeding algorithmically:* the use of an algorithm to make an approximate calculation and then to calculate the answer. For example, 8.3×11.2 can be calculated as $8 \times 11 = 88$, adding $0.2 \times 8 = 1.6$ and $0.3 \times 11 = 3.3$, making approximately $88 + 5 = 93$.

6.3. Variables or factors influencing computational estimation

Many variables or factors influencing the ability of computational estimation and the use of the strategies are cited in the literature. In the literature, in many cases, the terms factor or components are used; in this text we use the terms 'variable' or 'factor'. Based on the literature, the grouped presentation of the factors (Liu & Neber, 2012; Sowder & Wheeler, 1989; Van den Heuvel-Panhuizen, 2001) and our research experience, we propose the following taxonomy of variables influencing computational estimation and estimation in general. We adopt and use three main types of variables that influence and determine the flexibility/adaptivity determined by Verschaffel, Luwel, Torbeyns and Van Dooren (2009): (1) *Subject variables* in which are included variables concerning the *personal characteristics* of the solver, such as cognitive variables and knowledge, age, and emotional–affective variables. (2) *Task variables* in which are included variables related to *the situation of estimation* and *the characteristics of the problem*. In this group we can distinguish two subgroups: (2.1) variables related to *numbers* and *operations of the problem*, such as number magnitude and type of operation of the problem and (2.2) variables related to *the situation of estimation* and *the problem*, such as the context, (3) *Context variables*. Variables related to the sociocultural and environmental context.

6.3.1. Subject variables

6.3.1.1. Cognitive variables and knowledge

In a detailed analysis of the components involved in computational estimation carried by Sowder and Wheeler (1989, p. 132), they distinguished four sections of components: *I. Conceptual components, II. Skill components, III. Related concepts and skills and IV. Affective components*. These authors, in their Section III, on related concepts and skills, enumerate the following seven conceptual abilities and knowledges involved in computational estimation: *ability to work with powers of ten, knowledge of place value of numbers, ability to compare numbers by size, ability to compute mentally, knowledge of basic facts, knowledge of the properties of operations and their appropriate use* and *recognition that modifying numbers can change the outcome of computation*.

Van den Heuvel-Panhuizen (2001, pp. 200–201) describes the basic knowledge of numbers and operations that is essential for estimation and mentions that a great deal of this knowledge concerns mental arithmetic. This knowledge is: *Knowledge of numbers and measures, basic arithmetic facts, knowledge of operations* and *commutative, distributive and associative properties*.

In their research, Seethaler and Fuchs (2006) found that the computational estimation skills of third graders were predicted by cognitive factors such as arithmetic number combination skill, nonverbal reasoning, concept formation, working memory, and inattentive behaviour.

Star, Rittle-Johnson, Lynch and Perova (2009), in an experimental design study, examined the role of prior knowledge in the development of flexibility in the use of strategies in computational estimations. They found that students showing a high capacity for computational estimation in pre-test developed strategies that led to accurate estimates, and lower-ability students developed strategies that were easier to perform. Therefore, the role of prior knowledge is important for the progress of pupils to a higher level of computational estimation.

6.3.1.2. Age

Siegler and Booth (2005, p. 199) argue that the development of computational estimation begins surprisingly late and proceeds surprisingly slowly.

Computational estimation gradually improves over the third and fourth grades. Adults and students in sixth grade perform better than students in the fourth grade in the estimation of the sum of two- or three-digit addends (Lemaire & Lecacheur, 2002). Adults are correct more often than the students in the seventh grade, who are more correct than those in the sixth grade in estimating problems with multi-digit multiplications (LeFevre et al., 1993). The improvement in the speed of estimation of addition and multiplication is performed in a similar way to the improvement in accuracy at the same ages.

6.3.1.3. Affective or emotional variables

Self-efficacy and other emotional variables, such as intrinsic value and stress, may be related to the ability of computational estimation (Liu & Neber, 2012). Sowder and Wheeler (1989) highlighted other factors, such as confidence in the ability to do mathematics or estimation, tolerance of error and recognition of the usefulness of estimation. Tsao and Pan (2011) add the factors of self-confidence and tolerance of error and previous experience of the person with computational estimation, acceptability of computational estimation value and fun in studying computational estimation.

6.3.2. Task variables

6.3.2.1 Variables related to numbers and operations of the problem

The type of the number concerning roundness. Numbers that are close to a nice round number simplify and favour estimation after making obvious the need for rounding (Van den Heuvel-Panhuizen, 2001, p. 199).

The type of the number concerning the group of numbers (whole numbers, fractions, decimals, etc.). Fifth graders (Tsao & Pan, 2011) and high school students (Bana & Dolma, 2004) are more efficient in estimation problems with whole numbers than in problems involving decimals and fractions. Additionally, pre-service teachers had more difficulties in estimating problems with fractions than with decimal numbers (Tsao, 2013). For in-service teachers, the sum of the fractions is much more difficult than the sum of decimals and whole numbers in computational estimation problems (Lemonidis, Mouratoglou & Pnevmatikos, 2014).

The magnitude of numbers. Estimations with large numbers are usually more difficult, because they imply the calculation of zero and the relative size of the estimated result, especially in multiplication and division (Van den Heuvel-Panhuizen, 2001, pp. 199–200).

The quantity of numbers that must be rounded. Prospective teachers estimate the result of an addition with five terms, such as 35 + 42 + 40 + 38 + 44, more difficultly than addition with three terms, such as 1,378 + 236 + 442 (Lemonidis & Kaimakami, 2013).

The nature of the operations. In-service and pre-service teachers (Lemonidis & Kaimakami, 2013; Lemonidis, Mouratoglou & Pnevmatikos, 2014; Tsao, 2013), as well as high school students (Bana & Dolma, 2004) and undergraduate students (Hanson & Hogan, 2000), performed significantly better in addition and/or subtraction than in multiplication and division.

6.3.2.2 Variables related to the situation of estimation and the problem

Van den Heuvel-Panhuizen (2001, pp. 199–200) describes two factors related to the situation of estimation: *The desired degree of precision.* Estimations wherein a more global estimate is allowable and the answers can be less exact are easier for students. *The degree of openness of the problem.* Estimation problems with lacking or incomplete numerical data or problems where the students can independently decide about the rounding method and the degree of precision are more difficult.

Contextual and numerical problems. The existence or not of a context for the problem may be a significant factor for successful resolution. The results of research on this issue are contradictory. Researchers Goodman (1991) and Gliner (1991) found that for pre-service teachers, contextual problems are easier than numerical problems. On the other hand, Yang and Wu (2012), in a survey of 198 high school students, found that better performance occurred in numerical problems than the contextual problems. These contradictory results can be interpreted. Goodman (1991) comments that the context for estimation problems are more familiar to adults (p. 264), while Yang and Wu (2012) argue that contextual problems must be interpreted and translated into symbols and arithmetic operations and this conversion process confuses students.

Language of the problem. Imbo and LeFevre (2011), in a cross-cultural survey with 40 Belgian and 80 Chinese students, found that efficiency in computational estimation is associated with language. Participants who answer in a language other than their native language are slower in the process of solving computational estimation problems than in those answering in their native language.

Mode of problem presentation. Liu (2009) found that Chinese students performed better when tasks were presented visually than orally.

6.3.3. Context variables

6.3.3.1. Environmental variables

Liu and Neber (2012), in a cross-cultural study with Chinese and Polish sixth graders, found that the Chinese had better computational estimation and were only able to explain 61% of this variance. In terms of significant predictors, this percentage was explained by cognitive factors, such as complex-operand, three-operand and context-free, by motivational factors in maths and estimation, such as self-efficacy, intrinsic value, test anxiety, favourable attitude toward estimation, good and poor strategic tendencies, by environmental factors such as teacher use of estimation and country variable, and by interaction between teacher use of estimation and self-efficacy. They imply that the remaining 39% could be explained by exploring more environmental factors such as school and the orientation of the educational system towards mathematics and use of mathematics, more interactions between the environment and the person and between environments.

Imbo and LeFevre (2011) studied how cultural differences affect the estimation procedures. Participants (40 Belgian and 40 Chinese students) responded to numerical computational estimation problems either in their native language or in a language other than their native language (Chinese answered in English). They found that the Chinese students, although they are quicker and more accurate in calculations, were less flexible and adaptable in choosing the most appropriate estimation strategy than the non-Chinese. The authors conclude that the flexibility or difficulty in selecting appropriate strategies depends on the educational experiences of each cultural group and on tolerance towards estimations, because the Chinese preferred exact calculations.

According to the results of various surveys some of the above strategies are used more often than others. Rounding is the most commonly used strategy (Lemaire et al., 2000; LeFevre et al., 1993; Reys et al., 1982; Reys et al., 1991a) and compensation tends to be the last in frequency of use. For example, in a study by Lemaire et al. (2000) in order to estimate multi-digit addition, 64% of students in fifth grade used a rounding strategy and only 2% used compensation.

It was also found that students and adults, especially those skilled in estimation, know and use a variety of strategies in computational estimation. For example, when asked to estimate the sum of two three-digit numbers, French students in fifth grade used four main strategies: rounding with analysis, rounding without analysis, compensation and truncation (Lemaire et al., 2000). Almost all children (95%) used two strategies, 71% used at least three and 38% used all four.

It has been found that the age factor is important for the use of strategies. Various computational estimation strategies are used in different proportions by different age groups. For example, LeFevre et al. (1993) reported that children use *prior compensation*, e.g., calculating 75 × 200 to estimate 78 × 189, more frequently than *post compensation*, e.g., calculating 80 × 200 − (2 × 200), to estimate 78 × 189, while adults do the inverse. Students in the sixth grade use prior compensation strategies more often than the students in the eighth grade. In contrast, post compensation strategies are used equally in both age groups.

It has been found that computational estimation strategies vary in effectiveness. Efficiency of strategies improves with age. For example, Lemaire et al. (2000) find that when ten-year-olds were asked to estimate problems with three-digit addends, rounding with analysis (e.g., do 400 + 300 + 60 + 60 = 820 to estimate 459 + 356) produced a better estimation than the truncation (e.g., do 450 + 350 = 800), and the children were faster when they used truncation than rounding with analysis. All researchers comparing children of different ages find that estimation accuracy increases with age.

Finally, it was found that the use of various strategies depends on the characteristics of the problems. LeFevre et al. (1993) and Lemaire et al. (2000) observed that the selection of strategy in computational estimation is affected by the characteristics of the problem (such as the magnitude of numbers). In the research of LeFevre et al., rounding both numbers appear more frequently in problems with large numbers (e.g., 36 × 146) than problems with smaller numbers (e.g., 8 × 112).

In conclusion, we can say that, as in many other cognitive areas, in the research on the abilities of children in computational estimation strategies it is found that: (a) students use several strategies to estimate numerical problems, (b) these strategies vary in frequency and effectiveness and (c) the use and implementation of strategies in computational estimation change with age.

6.4. Research results in computational estimation

6.4.1. Research on students' abilities in computational estimation

Many studies show that students face difficulties in estimation (Levine, 1982; Sowder & Wheeler, 1989; Hanson & Hogan, 2000; Lemaire & Lecacheur, 2002; Siegler & Booth, 2004; Mildenhall et al., 2009). Researchers believe that these difficulties are connected with their failure to understand the meaning of the estimation procedure.

Sowder and Wheeler (1989) suggest that students make errors, usually of a conceptual nature. Some such errors show that students do not fully comprehend what estimation is, whereas other errors are the outcome of their inability to understand the procedure through which one can perform estimation. Adults display similar behaviour: for example, many prospective teachers do not know how to perform computational estimation in given problems and they perform accurate calculations using standard algorithms (Lemonidis & Kaimakami, 2013).

Computational estimation strategies have been investigated concerning the different levels of calculation ability of adults (Dowker, 1992; Dowker et al., 1996), children and teenagers (Levine, 1982; Baroody, 1989; Sowder & Wheeler, 1989; Reys et al., 1991a; Dowker, 1997; Lemaire et al., 2000), and target groups of different ages (LeFevre et al., 1993; Lemaire & Lecacheur, 2002). Different computational estimation strategies can be categorised according to different levels of generalisation.

Levine (1982) interviewed college students of varying mathematical backgrounds to identify strategies used to estimate. Each student was asked to estimate the answers to 20 questions. Levine found that poor estimators preferred to look for the exact computation and then rounded to find an estimate. She states that this method did not 'require the individual to sense any relationships or to have any "number sense" to carry it out' (p. 358). Good estimators, in Levine's study, used more strategies and appeared to be more flexible in their thinking than poor estimators. Yang (2005), in an interview setting, asked a series of questions to 21 sixth graders from four public schools in south Taiwan related to whole and decimal numbers, designed to assess their number sense. Rule-based methods, or no explanation, were the most popular responses from the students who answered correctly at each level, the low group (11 out of 15), the middle group (22 out of 29) and the high group (18 out of 26), according to the results (p. 321).

The outcome of these studies shows that quite early in the computational estimation developmental trajectory, children use a variety of strategies. These strategies are often naturally dependent on the operations of estimation given.

6.4.2. Research on teachers about computational estimation

Although professional mathematicians are very efficient with computational estimations (Alajmi, 2009; Dowker, 1992), the efficiency of both pre-service (Castro et al., 2002; Gliner, 1991; Goodman, 1991; Lemonidis & Kaimakami, 2013; Tsao, 2013; Yoshikawa, 1994) and in-service teachers (Alajmi, 2009; Dowker, 1992; Mildenhall et al., 2009; Tsao & Pan, 2013) in computational estimations is moderate or low. For instance, Castro et al., studied the difficulty of computational estimation tasks (with operations without context) in connection with operation type (multiplication and division) and number type (whole, decimal greater than one and decimal less than one) involved in them. An estimation test was administered to the teachers and some of them were selected to be interviewed. Castro et al. concluded that estimating with decimals less than one is more difficult for pre-service teachers than estimating with whole numbers or decimals greater than one. Most errors were produced in the estimation processes, due to the teachers' misconception of the operations of multiplication and division.

Additionally, Lemonidis and Kaimakami (2013) studied the performance, errors and computational estimation strategies used by 50 pre-service Greek teachers. They found that pre-service teachers were low-level estimators, facing more difficulties in multiplication and division estimation problems than in addition problems. Moreover, teachers were not familiar with and did not use strategies such as averaging and the compatible numbers strategy.

Tsao and Pan (2013) studied the understanding and the knowledge of practising teachers in Taipei in computational estimation and the instructional practices used by teachers in their everyday teaching practice. Six (three teachers with a mathematics/science major and three teachers with a non-mathematics/science major) fifth-grade elementary teachers participated in this study. The findings showed that all teachers were able to explain the meaning of computational estimation, and they efficiently used computational estimation strategies to solve problems. Their computational estimation strategies to solve problems included front-end, rounding, compatible number, special number, use of fractions, nice-number[1] and distributive property strategies. All six teachers used special numbers (1, 0, ½), and five of them used rounding and compatible number strategies. Four teachers used nice numbers while only one teacher used the front-end strategy and the distributive property. Alajmi (2009) examined 59 elementary and secondary mathematics teachers strategies for computational estimations in Kuwait. He found that, although some teachers were not aware of computational estimation, the majority of teachers used rounding in their computational estimations, while only 40% of strategies were used effectively, with 76% of those strategies being used by secondary teachers.

Scholars also investigated teachers' attitudes towards computational estimations and how these may affect everyday practice. Although teachers recognise the usefulness of computational estimation in daily life, there has not been general agreement on the integration of computational estimations in mathematics education. For instance, in Alajmi's (2009) study, although two-thirds of the teachers considered estimation to be an important life skill, nearly half of them do not consider computational estimation a significant topic in mathematics education. Tsao (2013) examined attitudes of 84 pre-service elementary teachers towards computational estimation in Southern Minnesota with the Computational Estimation Attitude Survey (CEAS) and their relations with their efficiency in computational estimations. The results showed a relationship between pre-service elementary teachers' computational estimation and their attitudes towards computational estimation; those who scored higher in computation estimations consider them as necessary, useful, and beneficial for life. Moreover, the relationship with mathematics is correlated with instructional practices (Tsao & Pan, 2013).

6.5. Research results of the School of Nature and Life Mathematics in computational estimation

6.5.1. Behaviours of fifth- and sixth-grade Greek students in computational estimation

Lemonidis, Nolka and Nikolantonakis (2014) examined behaviours in number sense, especially in computational estimation problems, of 596 Greek students of the fifth and sixth grades in primary school, who participated in a mathematics competition. The students who took part in this competition were self-selected. Their participation in the competition was completely voluntary. Their positive attitudes towards mathematics were perhaps their most distinctive feature. These students had not received any training in relation to computational estimation strategies before the test. Therefore, it was examined whether these students know how to perform computational estimation, and which strategies they use.

6.5.1.1. Examination procedure

This competition was held in May 2012 and took place in a time slot beyond the school timetable. It was a written examination. In each computational estimation problem, students were required to note their thought process and strategy for solving the problem.

6.5.1.2. The problems

The computational estimation problems were the following:

Fifth grade

Solve the above problems by mental computation and without using written operations. Explain the way you thought.

1.P.5. Mary ran 1/2 km in the morning and 3/8 km in the afternoon. Did she run at least 1 km?

2.P.5. A worker worked for 28 days earning €56 per day. How much money did he earn, approximately?

Sixth grade

1.P.6. In 816 ml of a substance, 9.84% is alcohol. How much alcohol is there approximately in the substance?

2.P.6. Give an approximate estimate of the sum of the following amounts of money:

€1.26,   €4.79,   €0.99,   €1.37,   €2.58.

6.5.1.3. Students' performance in computational estimation problems

In Table 6.1, the data show that a quite few students were familiar with and could perform computational estimation problems. Concerning students in the fifth grade, in the first and second problem, 28.3% and 22.3%, respectively were able to execute computational estimation correctly, while in the sixth grade, the percentage was higher, reaching 34.4% and 48.6% for each problem respectively. A considerably high percentage of the students attempted to solve problems employing exact calculation. In particular, fifth-grade students used exact calculation more than computational estimation in solving problems. More specifically, the first and the second problems were solved by means of employing exact calculation by 39.2% and 46.8% of the students. In the sixth grade, a small percentage of students (7.1%) solved the first problem accurately (1.P.6.) using exact calculation, as use of the written algorithm was difficult. Therefore, we came across a greater percentage of incorrect answers (46.1%) as well as failure to produce any answer (12.4%).

TABLE 6.1 Students' performance in computational estimation problems

	1.P.5. 1/2 plus 3/8	2.P.5. 28 x 56	1.P.6. 9.84% of 816	2.P.6. 1.26 + 4.79 + 0.99 + 1.37 + 2.58
Computational estimation	89(28.3%)	70(22.3%)	97(34.4%)	137(48.6%)
Exact calculation with algorithm	123(39.2%)	147(46.8%)	20(7.1%)	73(26%)
Wrong answer	82(26.1%)	93(29.6%)	130(46.1%)	48(17%)
No answer	20(6.4%)	4(1.3%)	35(12.4%)	24(8.5%)
Total	314(100%)	314(100%)	282(100%)	282(100%)

TABLE 6.2 Percentages of strategies used in correct computational estimation answers

Strategy	Description	1.P.5. 1/2 plus 3/8	2.P.5. 28×56	1.P.6. 9.84% of 816	2.P.6. $1.26 + 4.79 + 0.99 + 1.37 + 2.58$
Special numbers strategy	1/2 is half and 3/8 is less than the half 9.84% ≈ 10%	88(28%) **(99%)**★		59(20.9%) **(61%)**	
Rounding	$28 \times 56 \approx 30 \times 60 = 1800$ or rounding only one factor $1.3 + 4.8 + 1 + 1.4 + 2.6 \approx 11$		52(16.6%) **(74.5%)**		107(38%) **(78%)**
Rounding and compensation	$28 \times 56 \approx 30 \times 50$ $30 \times 60 - 100 = 1700$ $30 \times 56 - 112 = 1568$		17(5.4%) **(24.5%)**		3(1%) **(3%)**
Rounding and special numbers	$816 \approx 820$ and $9.84\% \approx 10\%$			33(11.7%) **(34%)**	
Front-end strategy or front-end and clustering	$1 + 4 + 1 + 1 + 2 = 9$ and $0.30 + 0.80 + 0.40 + 0.60 = 2.10$ $1 + 4 + 0 + 1 + 2 = 8$ and $0.5 \times 5 = 2.5$				22(7.8%) **(16%)**

★The percentage which corresponds to students who performed computational estimation correctly is marked in bold.

6.5.1.4. Strategies used by students

In the first problem (1.P.5.), less than 1/3 (28%) of students used the special number strategy, the most suitable strategy for this exercise. The rest of them used exact calculation by summing the fractions. In the second exercise (2.P.5.), among the few students who used rounding, (22%) were students (16.6%) who performed rounding of both terms: 28 × 56 → 30 × 60, while some of them (5.4%) performed rounding and compensation, which was the correction of rounding.

In problem 1.P.6., only one-third of students (34.4%) managed to produce the correct computational estimation. Most of these students (20.9%) used the special numbers strategy; they interpreted 9.84% as 10%. The rest of the students (11.7%) used the rounding of 816 ≈ 800 or 820 while simultaneously using a special number 9.84% ≈ 10%.

For almost half of the students (48.6%) who were able to perform a computational estimation in 2.P.6., most (38%) used rounding of addition terms (1.3 + 4.8 + 1 + 1.4 + 2.6 ≈ 11), whereas very few students (7.8%) used the front-end strategy, and added the integer parts (1 + 4 + 1 + 1 + 2 = 9) and the decimal parts (0.30 + 0.80 + 0.40 + 0.60 = 2.10) of the numbers separately.

We can assume that a strategy known and used by the students is that of rounding (38% in 2.P.6. or 16.6% in 2.P.5.). Very few students can actually use number-sense strategies or the most appropriate strategies for computational estimation, which are the special numbers strategy (28% in 1.P.5. and 20.9% in 1.P.6), the rounding and compensation strategy (5.4% in 2.P.5. and 1% in 2.P.6.) and the front-end strategy (7.8% in 2.P.6). Finally, we can say that students do not know number-sense strategies for estimation well.

6.5.2. In-service teachers' knowledge in computational estimation

In a survey undertaken by Lemonidis, Mouratoglou and Pnevmatikos (2014), the performance and strategies in computational estimation of 80 Greek in-service teachers of elementary school were examined. Participants executed 10 computational estimation problems with a paper-and-pencil procedure in the presence of the researcher. They were asked to write their on-going thoughts on the paper during their effort to solve the problem in order for these scripts to be informative about the strategies participants used to solve the problems. Additionally, ten of them were examined by interview.

6.5.2.1. Tasks

The 10 computational estimation problems given to our participants were designed to be solved by using one of five appropriate strategies, namely front-end strategy (P1 and P2), clustering or averaging (P5 and P6), rounding with post compensation (P9 and P10), compatible numbers strategy (P3 and P4) and special numbers strategy (P7 and P8). The following ten computational estimation problems were set for in-service teachers:

P1. Give an approximate estimate of the sum of the following amounts of money:

€1.26, €4.79, €0.99, €1.37, €2.58.

P2. Give an approximate number of students attending in all three schools:

secondary school A: 1,378 students, secondary school B: 236 students, secondary school C: 442 students.

P3. Six student groups prepared flower bouquets for the school feast. The groups prepared 27, 49, 38, 65, 56, 81 flower bouquets. How many flower bouquets have been prepared, approximately?

P4. A train of modern technology runs, 25,889 kilometres in 52 hours. How many kilometres does the train cover, approximately, in one hour?

P5. Is the following result approximately 200? 35 + 42 + 40 + 38 + 44.

P6. Six independent measurements were made by the team in order to find the height of Mount Everest: 28,990 ft, 28,991 ft, 28,994 ft, 28,998 ft, 29,001 ft, 29,026 ft.

Based on these measurements, what is the approximate height of Mount Everest?

P7. In 816 ml of a substance 9.84% is alcohol. How much alcohol is there, approximately, in the substance?

P8. Mary ran 1/2 km in the morning and 3/8 km in the afternoon. Did she run at least 1 km?

P9. A worker worked 28 days for €56 a day. How much will he be paid, approximately?

P10. A student who started skiing lessons completed 75 hours; the cost of each hour was €36. How much does he have to pay?

6.5.1.5. Strategies used in computational estimations

Although the most appropriate strategy for problems P1 and P2 is in accordance with the previous research about the front-end strategy, only 30% of teachers used this strategy properly in P1 and only 11.2% of teachers in P2, while the majority of participants used the rounding strategy (P1: 55% and P2: 80%). Similarly, although the most proper strategy for problems P3 and P4 is considered to be the most compatible numbers strategy, this strategy was used by only 22.5% of teachers in the problem P3 and 53.7% in the problem P4. Instead, in problem P3, the majority of teachers (66.3%) used the strategy of rounding. The strategy of accumulation or averaging is appropriate for the problems P5 and P6: 47.5% of teachers used this strategy properly for the problem P5, and 52.5% of teachers used it correctly for problem P6. The majority of teachers used the appropriate special numbers strategy to solve problems P7 and P8 (73.8% and 71.2% for the P7 and P8 problem respectively). For example, in problem P8 (1/2 km + 3/8 km), some teachers explained that 1/2 = 0.5 and 3/8 < 0.5 because 0.5 = 4/8, so the sum of fractions was not above 1 km. The most appropriate strategy for problems P9 and P10, namely the

TABLE 6.3 Percentages of strategies used in accurate answers

Strategies	Front-end	Rounding	Special numbers	Clustering or Averaging	Compatible numbers	Other
	$1+4+1+1+2=9$ and $0.30+0.80+0.40+0.60=2$ so $9+2=11$	$75 \times 36 \approx 80 \times 40 = 3{,}200$ with compensation	1/2 is a half, 3/8 is less than a half	$35+42+40+38+44$ all are close to 40	$27+81 \approx 100$ $38+65 \approx 100$ $49+56 \approx 100$ total 300	
Problem						
P1	**30%★**	55%			7.5%	1.3%
P2	**11.2%★**	80%			1.3%	1.3%
P3	2.5%	66.3%			**22.5%★**	
P4		30%			**53.7%★**	2.5%
P5	2.5%	28.7%		**47.5%★**	5%	1.3%
P6		17.5%		**52.5%★**	1.3%	6.2%
P7			**73.8%★**			
P8			**71.2%★**			1.3%
P9	5%	**57.5%★**			3.7%	1.3%
P10	10%	**42.5%★**				1.3%

Note: Percentages in bold indicate the appropriate strategy for each problem, according to the literature review.

rounding strategy, was used by a satisfactory rate of teachers (57.5% and 42.5% for P9 and P10 problem, respectively). For example, in problem P10 (75 × 36), some teachers rounded the number 36 ≅ 40 up and rounded the number 76 ≅ 70 down. After this, they multiplied 40 × 70 and concluded on 2,800.

6.6. Instruction and instructional interventions

The improvement of computational estimation capacity with age, which we mentioned above, is confirmed by modern psychological studies. Lemaire and Lecacheur (2011), in a survey of third, fifth and seventh graders, studied the relation of computational estimation capability with executive functions. Students selected the best strategy (rounding up or rounding down) for estimating answers to two-digit addition problems. To measure participants' executive functions they used known measuring tools (Stroop, ELFT and TMT).

The results were consistent with the findings of previous studies, since the authors found that speed and accuracy of children in computational estimation problems improve with age. Strategy selection also improved with age. Increased efficiency in executive functions was associated positively (but without a causal relationship) with age-related improvement in children's skill in strategy selection. Finally, results showed that two additional executive functions, inhibition processes and cognitive flexibility, are related to strategy selection and to age-related differences in strategy selection.

According to the preceding, concerning the development of understanding in computational estimation, a general conclusion may be that we should not rush to teach estimation from an early age, teaching young students various strategies in a superficial way.

In the curriculum of estimation we should take into account spontaneous beliefs and strategies of children and their development as well. Students should become familiar with various contexts and situations of everyday life where estimation is present, not only in order to be able to appreciate its usefulness, but also to develop a deep and complex understanding.

According to Reys (1986, p. 1), estimation, like problem solving, is associated with a variety of abilities and improves over a long period of time. Like problem solving, estimation cannot be taught as a simple concept in a single lesson. It should permeate many areas of the curriculum and be developed effectively by encouraging the study of various areas of mathematics. When taught as separate content, the efforts of students may be unproductive and instruction may leave them with a general discontent and distrust in a multitude of strategies. To be truly effective, estimation should be completely incorporated into teaching, with the simultaneous development of several areas (Reys, 1986, p.1):

1. an awareness for and an appreciation of estimation;
2. number sense;
3. number concepts;
4. estimation strategies.

Van den Heuvel-Panhuizen (2001, pp. 177–178) presents the learning–teaching trajectory in estimation, followed by the children, subdivided into four sub-domains:

- rounding off numbers
- estimations in addition and subtraction
- estimations in multiplication and division
- estimation with incomplete data.

The author notes that the four phases are interwoven with each other and continuously involve each other. In every phase, there is an evolution from an informal phase to a rule-directed phase and finally to a flexible phase. For example, in the situations of estimations with addition, subtraction, multiplication and division in informal estimation phase the students determine answers globally without using the standard rounding off rule. In the rule-directed estimation phase, students apply the standard rounding off rule and in the flexible estimation phase the students apply more balanced estimation methods.

Some experimental research interventions have been performed aimed at the instruction of computational estimation. The study of Bobis (1991) was done at 15 weeks experimental teaching in 101 pupils of fifth grade. The results indicated a significant improvement of the experimental group in the performance of computational estimation, and no significant improvement by the control group. Experimental instruction led to the implementation of valid estimation strategies and to the improvement of pupils' capacity to explain.

Mildenhall et al. (2009) published a study of a professional learning intervention with one teacher. The results highlight the need for training teachers to develop positive attitudes towards computational estimation (for example, recognition of its value and necessity) and its integration into the teaching practice of primary schools.

Star, Rittle-Johnson, Lynch and Perova (2009) examined the role of students' prior knowledge of estimation strategies in the effectiveness of interventions designed to promote strategy flexibility. Their results indicated that prior knowledge affects the strategies that students learn. High-ability students learned strategies that led to more accurate estimates, while students with less fluency adopted strategies that were easy to implement.

Star and Rittle-Johnson (2009) ran an experimental instruction where 157 students in the fifth and sixth grades learned about estimation either by comparing alternative solution strategies or by reflecting on strategies one at a time. Finally, all improved their procedural knowledge (knowledge of computational estimation strategies). Instruction by strategy comparison resulted in greater flexibility. Also, pupils with some prior knowledge of assessment strategies gained a better conceptual understanding.

Mildenhall and Hackling (2012) investigated the impact of professional learning intervention on the teaching and learning of computational estimation. It appears that professional learning intervention did not impact students' general beliefs about mathematics, as they perceived estimation work as something outside normal school mathematics. The students' performance of computational estimation and their use of reasoned estimation strategies were enhanced as a result of their teacher's involvement in professional learning intervention.

Note

1 Some authors use *known* or *nice-number* strategies, which are a kind of rounding strategy. We round the numbers in the operation so as to have known numbers to calculate, e.g., to estimate the quotient of 6,470 ÷ 73, we estimate that 6,300÷70 is about 90.

References

Alajmi, A. H. (2009). Addressing computational estimation in the Kuwaiti curriculum: Teachers' views. *Journal of Mathematics Teacher Education, 12*(4), 263–283. DOI: 10.1007/s10857-009-9106-3.

Anestakis, P. & Lemonidis, Ch. (2014). Computational estimation in an adult secondary school: A teaching experiment. *MENON: Journal Of Educational Research, 1st Thematic Issue,* 28–45.

Bana, J. & Dolma, P. (2004). The relationship between the estimation and computation abilities of Year 7 students. In I. Putt, R. Faragher & M. McLean (Eds.), *Proceedings of the 27th annual conference of the Mathematic Education Research Group of Australasia* (Vol. 1, pp. 63–70). Townsville: MERGA.

Barroody, A. J. (1989). Kindergartners' mental addition with single-digit combinations. *Journal for Research in Mathematics Education, 20,* 159–172.

Bobis, J. (1991). The effect of instruction on the development of computational estimation strategies. *Mathematics Education Research Journal, 3*(1), 7–29.

Castro, C., Castro, E., & Sergovia, I. (2002). Influence of number type and analysis of errors in computational estimations tasks. In A. D. Cockburn & E. Nardi (Eds.), *Proceedings of the 26th Conference of the International Group for the Psychology of Mathematics Education: Vol.2* (pp. 201–208). Norwick, UK.

Dowker, A. (1992). Computational estimation strategies of professional mathematicians. *Journal for Research in Mathematics Education, 23*(1), 45–55.

Dowker, A. (1997). Young children's addition estimates. *Mathematical Cognition, 3,* 141–154.

Dowker, A., Flood, A., Griffiths, H., Harriss, L. & Hook, L. (1996). Estimation strategies of four groups. *Mathematical Cognition, 2,* 113–135.

Faulkner, V. (2009). The components of number sense: An instructional model for teachers. *Teaching Exceptional Children, 41*(5), 24–30.

Gliner, G. S. (1991). Factors contributing to success in mathematical estimation in preservice teachers: Types of problems and previous mathematical experience. *Educational Studies in Mathematics, 22*(6), 595–606.

Goodman, T. (1991). Computational estimation skills of pre-service elementary teachers. *International Journal of Mathematical Education in Science and Technology, 22,* 259–272.

Greeno, J. (1991). Number sense as situated knowing in a conceptual domain. *Journal for Research in Mathematics Education, 22*(3), 170–218.

Hanson, S. & Hogan, Th. (2000). Computational estimation skill of college students. *Journal for Research in Mathematics Education, 31*(4), 483–499.

Imbo, I. & LeFevre, J.-A. (2011). Cultural differences in strategic behaviour: A study in computational estimation. *Journal of Experimental Psychology: Learning, Memory, and Cognition, 37*(5), 1294–1301.

LeFevre, J. A., Greenham, S. L. & Waheed, N. (1993). The development of procedural and conceptual knowledge in computational estimation. *Cognition and Instruction*, *11*, 95–132.

Lemaire, P., Arnaud, L. & Lecacheur, M. (2004). Adults' age-related differences in adaptivity of strategy choices: Evidence from computational estimation. *Psychology and Aging*, *19*(3), 467–481.

Lemaire, P. & Lecacheur, M. (2002). Children's strategies in computational estimation. *Journal of Experimental Child Psychology*, *82*, 281–304.

Lemaire, P. & Lecacheur, M. (2011). Age-related changes in children's executive functions and strategy selection: A study in computational estimation. *Cognitive Development*, *26*, 282–294.

Lemaire, P., Lecacheur, M. & Farioli, F. (2000). Children's strategy use in computational estimation. *Canadian Journal of Experimental Psychology*, *54*(2), 141–148.

Lemonidis, Ch. & Kaimakami, A. (2013). Prospective elementary teachers' knowledge in computational estimation. *Menon: Journal of Educational Research*, *22b*, 86–98.

Lemonidis, Ch., Mouratoglou, A. & Pnevmatikos, D. (2014). Elementary teachers' efficiency in computational estimation problems. *MENON: Journal Of Educational Research*, *1st Thematic Issue*, 144–158.

Lemonidis, Ch., Nolka, E. & Nikolantonakis, K. (2014). Students' behaviours in computational estimation correlated with their problem-solving ability. *MENON: Journal Of Educational Research*, *1st Thematic Issue*, 46–60.

Levine, D. R. (1982). Strategy use and estimation ability of college students. *Journal for Research in Mathematics Education*, *13*, 350–359.

Liu, F. (2009). Computational estimation performance on whole-number multiplication by third- and fifth-grade Chinese students. *School Science and Mathematics*, *109*(6), 325–337.

Liu, W. & Neber, H. (2012). Estimation skills of Chinese and Polish grade 6 students on pure fraction tasks. *Journal of mathematics education*, *5*(1), 1–14.

McIntosh, A. (2004). Where we are today. In A. McIntosh & L. Sparrow (Eds.), *Beyond written computation* (pp. 3–14). Perth: MASTEC.

Mildenhall, P. (2009). A study of teachers' learning and teaching of computational estimation: Getting started. In B. Kissane, M. Kemp, L. Sparrow, C. Hurst & T. Spencer (Eds.), *Mathematics it's mine: Proceedings of the 22nd biennial conference of The Australian Association of Mathematics Teachers* (pp. 153–158). Fremantle: AAMT.

Mildenhall, P. & Hackling, M. (2012). The impact of a professional learning intervention designed to enhance year six students' computational estimation performance. In J. Dindyal, L. P. Cheng & S. F. Ng (Eds.), *Mathematics education: Expanding horizons (Proceedings of the 35th annual conference of the Mathematics Education Research Group of Australasia)*. Singapore: MERGA.

Mildenhall, P., Hackling, M., & Swan, P. (2009). Computational estimation in the primary school: A single case study of one teacher's involvement in a professional learning intervention. In L. Sparrow, B. Kissane & C. Hurst (Eds.), *Shaping the future of mathematics education: Proceedings of the 33rd annual conference of the Mathematics Education Research Group of Australasia*. Fremantle: MERGA.

National Council of Teachers of Mathematics (NCTM) (2000). *Principles and standards for school mathematics*. Reston, Va.: Author.

Reys, B. J. (1986). Teaching computational estimation: Concepts and strategies. In H. L. Schoen & M. J. Zweng (Eds.), *Estimation & Mental Computation, 1986 yearbook, National Council of Teachers of Mathematics* (pp. 31–44). Reston, Va.: National Council of Teachers of Mathematics.

Reys, B. J., Reys, R. W. & Penafiel, A. F. (1991a). Estimation performance and strategy use of Mexican fifth- and eighth-grade student sample. *Educational Studies in Mathematics*, *22*, 353–375.

Reys, R. E. (1984). Mental computation and estimation: Past, present and future. *Elementary School Journal*, *84*(5), 546–557.

Reys, R. E., Bestgen, B. J., Rybolt, J. F. & Wyatt, J. W. (1982). Processes used by good computational estimators. *Journal for Research in Mathematics Education*, *13*(3), 183–201.

Reys, R. W., Reys, B. J., Nohda, N. & Ishida, J. (1991b). Computational estimation performance and strategies used by fifth- and eighth-grade Japanese students. *Journal for Research in Mathematics Education*, *22*(1), 39–58.

Schoenfeld, A. H. (1985). *Mathematical problem solving*. Orlando, FL: Academic Press.

Seethaler, P. M. & Fuchs, L. S. (2006). The cognitive correlates of computational estimation skill among third grade students. *Learning Disabilities Research & Practice*, *21*(4), 233–243.

Segovia, I. & Castro, E. (2009). Computational and measurement estimation: Curriculum foundations and research carried out at the University of Granada, Mathematics Didactics Department. *Electronic Journal of Research in Educational Psychology*, *17*(7) 499–536.

Siegler, R. S. & Booth, J. L. (2004). Development of numerical estimation in young children. *Child Development*, *75*(2), 428–444.

Siegler, R. S. & Booth, J. L. (2005). Development of numerical estimation: A review. In J. I. D. Campbell (Ed.), *Handbook of mathematical cognition* (pp. 197–212). New York: Psychology Press.

Sowder, J. T. (1992). Estimation and number sense. In D. Grouws (Ed.), *Handbook of research on mathematics teaching and learning* (pp. 371–389). New York: Macmillan.

Sowder, J. T. & Wheeler, M. M. (1989). The development of concepts and strategies used in computational estimation. *Journal for Research in Mathematics Education*, *20*, 130–146.

Star, J. R. & Rittle-Johnson, B. (2009). It pays to compare: An experimental study on computational estimation. *Journal of Experimental Child Psychology*, *102*, 408–426.

Star, J. R., Rittle-Johnson, B., Lynch, K. & Perova, N. (2009). The role of prior knowledge in the development of strategy flexibility: The case of computational estimation. *ZDM*, *41*(5), 569–579.

Tsao, Y. L. (2013). Computational estimation and computational estimation attitudes of pre-service elementary teachers. *US-China Education Review B*, *3*(11), 835–846.

Tsao, Y. L. & Pan, T. R. (2011). Study on the computational estimation performance and computational estimation attitude of elementary school fifth graders in Taiwan. *US-China Education Review*, *8*(3), 264–275.

Tsao, Y. L. & Pan, T. R. (2013). The computational estimation and instructional perspectives of elementary school teachers. *Journal of Instructional Pedagogies*, *11*, 1–15.

Van den Heuvel-Panhuizen, M. (2001). *Children learn mathematics: A learning–teaching trajectory with intermediate attainment targets for calculation with whole numbers in primary school*. Utrecht: Freudenthal Institute & National Institute for Curriculum Development.

Verschaffel, L., Luwel, K., Torbeyns, J. & Van Dooren, W. (2009). Conceptualizing, investigating, and enhancing adaptive expertise in elementary mathematics education. *European Journal of Psychology of Education*, *24*(3), 335–359.

Yang, D. C. (2005). Number-sense strategies used by 6th grade students in Taiwan. *Educational Studies*, *31*(3), 317–333.

Yang, D. C. & Wu, S. S. (2012). Examining the differences of the 8th-Graders' estimation performance between contextual and numerical problems. *US-China Education Review A*, *12*, 1061–1067.

Yoshikawa, S. (1994). Computational estimation: Curriculum and instructional issues from the Japanese perspective. In R. Reys & N. Nohda (Eds.), *Computational alternatives for the twenty-first century: Cross-cultural perspectives from Japan and the United States* (pp. 51–62). Reston, VA: National Council of Teachers of Mathematics.

7
LEARNING DIFFICULTIES AND MENTAL CALCULATION

This chapter will attempt to outline the main features of mental calculation for students with learning difficulties in mathematics. Initially, it will be demonstrated how learning difficulties in mathematics are defined in the literature. Secondly, the function of memory will be explored, with further emphasis on working memory and the brain in relation to mental calculations. Then different strategies of mental calculation and the performance of pupils with learning difficulties in mathematics will be presented. Finally, research will be presented related to instructional interventions and proposals for tackling the problems of these children in mental calculations.

7.1. Difficulties in mathematics and individual differences: definition of terms

Research in the UK has indicated that low level numeracy or mathematical literacy, which negatively affects citizens in economic, social and psychological ways, has a great cost to the state. Difficulties with numeracy as well as other factors were also associated with reduced employment opportunities, school dropout and exclusion from school. In the UK, almost 25% of adults have poor functional numeracy (Bynner & Parsons, 2005; Gross, 2009). The factors that affected poor performance in mathematics, apart from formal education problems (e.g., teaching content, teaching methods, teaching materials, etc.), may be exogenous, such as low socioeconomic level, membership in an ethnic minority, negative family environment, etc., but there are also endogenous factors, associated with the individual characteristics and peculiarities of each student. The American Psychiatric Association (1994) gives the following definition for difficulties in mathematics:

> **Mathematics disability**: The child must substantially underachieve on a standardized test relative to the level expected given age, education and intelligence and must experience disruption to academic achievement or daily living.

Research results showed that 6.8% of students consistently failed at math, despite having normal intelligence, living in a normal environment, having the same opportunities at school learning as their peers and showing no sensory or serious emotional problems (Kosc, 1974; Badian, 1983; Geary, 1994). Similar disorders in learning may be attributed to glitches or peculiarities of basic psychological functions or the central nervous system. These disorders of learning are termed: *(special) learning difficulties*. Regarding mathematics, the World Health Organization (1994) gives the following definition for the specific disorder of arithmetical skills:

> **Specific disorder of arithmetical skills:** Specific impairment in arithmetic skills that is not solely explicable on the basis of general mental retardation or of inadequate schooling.

In the area of mathematical difficulties, two distinct learning disorders seem to be common, namely *dyslexia* and *dyscalculia*. Dyslexia is a learning difficulty that primarily affects skills related to the ability to correctly and fluently read and spell words. There are a large number of different definitions and descriptions of dyslexia, which may be variously suitable for certain purposes and contexts. In October 2007, the British Dyslexia Association Management Board approved the following definition:

> **Dyslexia:** Dyslexia is a specific learning difficulty that mainly affects the development of literacy and language related skills. It is likely to be present at birth and to be lifelong in its effects. It is characterised by difficulties with phonological processing, rapid naming, working memory, processing speed, and the automatic development of skills that may not match up to an individual's other cognitive abilities.
>
> It tends to be resistant to conventional teaching methods, but its effect can be mitigated by appropriately specific intervention, including the application of information technology and supportive counselling.

Dyscalculia is associated with severe learning difficulties and use of mathematical concepts and procedures. A definition of dyscalculia is given below (DfES, 2001):

> **Dyscalculia:** A condition that affects the ability to acquire arithmetic skills. Dyscalculic learners can have difficulty understanding simple number concepts, lack an intuitive grasp of numbers, and have problems learning number facts and procedures. Even if they produce a correct answer or use a correct method, they might do so mechanically and without confidence.

There is also the term *acalculia*, different from dyscalculia, although some authors have used the terms interchangeably. Henschen (1925) proposed the term acalculia. He then defined acalculia as an impairment in computational skills resulting from brain injury. Patients with acalculia have difficulty performing simple mathematical tasks, such as adding, subtracting, and even simple situations of comparing two numbers. Acalculia is acquired later in life due to neurological injury such as stroke, while dyscalculia is a specific developmental disorder first observed during the acquisition of mathematical knowledge.

Moreover, another term worth mentioning is developmental dyscalculia. Kosc (1970, p. 192) proposed the following definition of *developmental dyscalculia*:

> **Developmental dyscalculia** is a structural disorder of mathematical abilities which has its origin in a genetic or congenital disorder of those parts of the brain that are the direct anatomico-physiological substrate of the maturation of mathematical abilities adequate to age, without a simultaneous disorder of general mental functions.

It can be seen that the two previous terms, *developmental dyscalculia* and *acquired dyscalculia*, seem to refer to the same condition. However, there is a clear distinction between them in the literature. Acquired dyscalculia concerns people who had learned mathematics, but later, in childhood or adolescence, or more often in adulthood, lost this ability, due to some acquired disorder associated with brain damage due to an accident or other cause. The term developmental dyscalculia regards people of school age at their first coming into contact and acquiring mathematical knowledge and skills (Gilbert, 1992).

Recently, some authors (e.g., Price & Ansari, 2013) have made the distinction between primary and secondary developmental dyscalculia. On one hand, *primary developmental dyscalculia* concerns children with severe math difficulties. This developmental learning disorder undermines children's ability to process basic numerical magnitude information, which in turn undermines the acquisition of school-level arithmetic skills. On the other hand, *secondary developmental dyscalculia* refers to mathematical deficits stemming from external factors such as poor teaching, low socioeconomic status, and behavioural attention problems or domain-general cognitive deficits (Price & Ansari, 2013, p. 11).

In addition to the above distinction, there are numerous divisions of dyscalculia in the literature. Geary (2000, 2003), for instance, proposed the following classification of mathematical difficulties:

1. *Semantic memory*. This subcategory includes children who have difficulty accessing and retrieving mathematical facts from memory, for example answers to simple arithmetic problems or operations. Errors appear frequent, although the retrieval reaction time of correct answers may vary considerably.
2. *Procedural memory*. Procedural memory concerns children who have difficulty understanding and applying mathematical procedures. Characteristically,

these children struggle when using their working memory and they also have difficulties in counting.
3. *Visuo-spatial memory*. This last subcategory concerns children who have difficulty understanding spatially represented numerical information, such as misalignment of columns. In addition, they seem to make place value errors and struggle with geometry.

In essence, the fact that the dyslexic population can be categorised into different subgroups of dyscalculia complicates diagnosis, as well as treatment. This complication can lead to confusion, because learning disabilities might be the foundation of mathematical difficulties, as was previously mentioned.

7.1.1. Relation between dyscalculia and dyslexia

Dyscalculia has still not been recognised by many scholars as an independent syndrome in the field of learning disabilities. There are three main trends of the relationship between dyslexia and dyscalculia. The first trend argues that dyslexia is a broad term that includes the symptoms of dyscalculia. Scientists such as Miles (1992) and Chinn and Ashcroft (1993) state that difficulties in mathematics have a linguistic origin. The second trend supports the idea that dyscalculia is different from dyslexia and is a distinct special learning difficulty that sometimes appears in dyslexics. The third trend describes dyscalculia as nothing more than failure to learn basic math skills due to weak memory problems or simply poor teaching. For the purposes of this book, I will work with the second trend, which considers the syndrome of dyscalculia to be separate from dyslexia.

From this point of view, dyscalculia often coexists with other learning difficulties (e.g., dyslexia) and rarely appears on its own in mathematics. Moreover, dyscalculia can affect the whole range of mathematical knowledge and skills or just one specific area, such as arithmetical operations or problem solving.

7.2. How does memory function?

There are many research studies concerning the nature and characteristics of memory (e.g., Baddeley, 1986, 1990; Parkin, 1993). This section will briefly outline the most important features of memory associated with mental calculations.

Memory refers to the mechanisms of the brain that retain and retrieve information from past experiences. More specifically, there are three basic memory functions: *coding, retention (storage)* and *retrieval* of information. Each process represents a different stage of mnemonic function. Specifically, encoding refers to the process with which the individual forms a mental representation of sensory data. Then the encoded information is memorised: this action is called retention. Finally, during retrieval, a person retrieves and uses information stored in memory.

Atkinson and Shiffrin (1968) proposed a model of memory consisting of three structural types of memory storage, namely:

a. *sensory resister*, which holds a relatively small amount of information for a very short time;
b. *short-term storage*, which also holds information for a short time, and has a relatively low retention capacity (approximately seven items); additionally the items shown in this memory disappear after a few seconds; and
c. *long-term storage*, which has a seemingly unlimited capacity and retention time information and can last from a few hours up to the whole life of the individual.

Another model of memory, originally proposed by Craick and Lockhart (1972), is the depth or level of cognitive information processing. According to this model, memory consists of three or any particular number of discrete restraint systems, but the quality of the mnemonic retention varies according to the 'depth' or the level of coding (processing) of the existing information. The retention level of the information depends largely on the way in which these elements are coded. The deeper the level of cognitive processing of information is, the greater the likelihood of information to be retrieved.

Psychologists have also divided long-term memory into two major categories: *episodic memory* and *semantic memory*. Episodic memory includes memories of particular observed or experienced events. Individuals usually have an approximate idea of when and where these events occurred. Semantic memory concerns general factual knowledge about human life, for example, knowledge of words, people, school experiences and information, etc. Facts concerning arithmetic, numerical facts and procedures are held in semantic memory.

According to the above, we could assume that long-term semantic memory has greater impact on arithmetic. This memory is apparently considered more important for arithmetic. However, many researchers believe that short-term memory also plays a very important role in the arithmetic. This is because short-term memory does not only serve as storage, but acts as *working memory*. Working memory will be further analysed in the following section.

7.2.1. Working memory

The term working memory refers to a mental workspace in which information may be stored and processed for the short periods of time required by cognitive activities. One of the most popular models, used in many studies, is the working memory model proposed by Baddeley and Hitch (1974, see also Baddeley, 1996, 2000). According to this model, working memory consists of a number of separated but temporarily intertwined memory systems.

The central *executive system* is a system of limited processing capacity, which is the core of this model of working memory. It regulates the flow of information through working memory. The central executive system also coordinates access and recall of the most stable systems of knowledge, such as long-term memory, controlling action and scheduling multiple cognitive activities.

In this model of working memory, the central executive system is supplemented by two 'slave' systems, specialising in maintaining information in special informational structures:

a. The *articulatory or phonological loop system* is a caching mechanism capable of storing limited information of phonological form. It holds inner speech and enables verbal comprehension and phonological repetition (without which the auditory information is 'off' after 2 seconds).
b. The *visuo-spatial sketchpad* is a system capable of preserving visual images for a short period of time.

Recent characterisations of the central executive system (e.g., Baddeley, 1996; Miyake, Friedman, Emerson, Witzki, Howerter & Wager, 2000) support the idea of *distinct executive functions*. These distinct executive functions include *inhibition* (suppression of strong but non goal-appropriate behaviour), *shifting* (disengagement of an irrelevant task set or an inappropriate strategy and subsequent activation of one more appropriate) and *updating* (codification and evaluation of incoming information on their relevance to the operable state and subsequent revision of the information stored in memory).

Baddeley (2000) recently identified an additional component of working memory, the *episodic buffer*. The episodic buffer is responsible for integrating information from various sources into the cognitive system, as well as into the two previously mentioned memory systems, temporary and long-term memory. This system has limited holding capacity and is responsible for the integration of information from various auxiliary systems and long-term memory representations into single episodes. In other words, the system connects information from various systems of working memory to achieve a level of consistency. For the purposes of this chapter the detailed structure of the episodic buffer must be analysed in accordance with assessment methods.

To accomplish this, it must to be made clear that working memory is not uniform. It includes several processes such as holding information in short-term memory, using this information to guide actions and monitoring the number of steps during the solution of a problem or when performing an activity. Finally, working memory is also capable of overseeing the results of one of these steps while performing the others.

It is very important to monitor relevant information during the solution of a problem and especially to monitor each step in the solution of the problem, while the next step is being planned or carried out. For instance, in mental calculations, it is important to monitor units while adding tens, or to monitor the sense within a word problem while at the same time handling the numbers. The monitoring of steps in arithmetic requiring multiple steps is likely to be a serious problem for people with impaired memory.

Hence, long-term memory may be particularly important in memory function for number facts as well as to remember the way in which numerical procedures should be made. Working memory, therefore, seems to be particularly important for placing the right procedures in the correct order, without deviating or getting confused (Dowker, 2005).

7.3. The brain, arithmetic and mental calculations

In recent decades, much research has been developed on the brains of patients and healthy individuals in areas and networks that are particularly important for arithmetic. This section will attempt to present research results on the regions of the brain that determine a person's behaviour concerning arithmetic, especially in areas related to mental calculation. Additionally, it will be demonstrated how individual differences in brain function play a role in arithmetic and especially in mental calculations.

Studies of patients facing difficulties in some or all areas of arithmetic showed arithmetic consists of many components. Moreover, there may be individual differences in the various operating levels of these different components. This synthetic nature of arithmetic was identified in the research as a phenomenon usually called *dissociations*, namely through failures in one component of arithmetic, while another is kept in good condition (e.g., one can add but not subtract).

Brain studies are usually performed through the method of *Magnetic Resonance Imaging* (MRI). Most studies of the brain concerned adults who suffered illnesses or accidents such as bumps that caused a brain injury. Children's difficulties in arithmetic may sometimes be derived from such faults or weaknesses of the brain. Thus, some children face numerical difficulties due to specific genetic conditions, such as Turner syndrome, Williams syndrome or Fragile X syndrome (Butterworth, Grana, Piazza, Girelli, Price & Skuse, 1999; Mazzocoo, 2001). These genetic characteristics can mean that numerical difficulties are associated with unusual features of the brain. Furthermore, Shalev, Manor, Kerem, Ayali, Badichi, Friedlander and Gross-Tsur (2001) found that developmental dyscalculia, even when not associated with a known genetic disorder, has a strong tendency to be inherited.

In recent years, by virtue of brain lesion studies, fMRI studies were performed in healthy individuals. The results show that different numerical concepts activate different brain regions. For instance, numerical concepts generally activate the *fronto-parietal* brain regions and *basal ganglia*. The *left parietal cortex* is activated during exact calculations, while estimated and approximate calculations activate the right hemisphere more (Dehaene, Tzourio, Frank, Raynaud, Cohen, Mehler & Mazoyer, 1996).

Several data of functional neuroimaging researches indicate the important role that the regions of the *parietal cortex* play in mental calculations. (Dehaene, Molko, Cohen & Wilson, 2004; Dehaene, Piazza, Pinel & Cohen, 2003; Dehaene, Spelke, Pinel, Stanescu & Tsivkin, 1999; Dehaene et al., 1996; Delazer, Domahs, Bartha, Brenneis, Lochy, Trieb & Benke, 2003; Delazer, Ischebeck, Domahs, Zamarian, Koppelstaetter, Siedentopf, Kaufmann, Benke & Felber, 2005; Ischebeck, Zamarian, Egger, Schocke & Delazer, 2007; Ischebeck, Zamarian, Siedentopf, Koppelstatter, Benke, Felber & Delazer, 2006; Menon, Rivera, White, Eliez, Glover & Reiss,

2000; Menon, Rivera, White, Glover & Reiss, 2000; Venkatraman, Siong, Chee & Ansari, 2006; Zago, Pesenti, Mellet, Crivello, Mazoyer & Tzourio, 2001; Zago & Tzourio-Mazoyer, 2002). In addition, the results of many studies indicate that the left parietal cortex is particularly important for verbal recall of known arithmetic facts, such as multiplication tables, while other areas of the brain play a greater role in situations that require more arithmetic reasoning (Cohen, Dehaene, Chochon, Lehricy & Naccache, 2000; Kazui, Kitagaki & Mori, 2000).

Roland and Friberg (1985), in the first neuroimaging study of arithmetic thinking, concluded that the angular gyrus is involved in the retrieval of the memory for subtraction and integers. Similarly, Rueckert, Lange, Partiot, Appollonio, Litvan, Le Bihan and Grafman (1996) examined the brain areas involved in mental subtractions and found that the areas of the left parietal cortex angular and supramarginal gyrus are involved in the solution of arithmetic problems. Besides these parietal areas, prefrontal cortices are also affected by the subtractions. Dehaene et al. (1999) discovered the special role played in calculation by the left parietal cortex, especially the angular gyrus. During exact calculations, they found greater activation of the angular gyrus relative to approximate calculations. Moreover, they attribute an important role of the angular gyrus in the exact calculation of arithmetic facts.

However, special attention should be paid to the assumptions made regarding the process required in any proposed arithmetic operation. This is indicated because several different strategies may be used for the same task of different individuals or from the same individual at different times (Lefevre et al., 2003, Siegler & Jenkins, 1989). Moreover, it is well known that arithmetical performance is not based purely on arithmetical abilities, but also on non-arithmetical abilities. Non-arithmetical abilities mainly refer to linguistic abilities, spatial abilities, memory, attention, etc. Thus, differences in brain function from arithmetic stimuli can be derived both from differences in brain function that are specific to the arithmetic and by differences in the function of brain areas associated with non-arithmetical abilities but related to the arithmetical performance.

As has previously been noted, parietal circuits are involved in mental calculations. Furthermore, developmental studies have shown that the activation of these areas is not static, but changes according to the training and development of the individual. Recent research, therefore, demonstrates that the activation of these parietal circuits depends not only on the cognitive demand or task under investigation. However, the activation of these regions is also modulated in accordance with the development and training that can show the dynamic changes in dependence on one of these circuits to the other (Grabner, Ansari, Reishofer, Stern, Ebner & Neuper, 2007, p. 347).

7.4. Number facts and memory

Children's acquisition of arithmetic facts, their retrieval from memory and their use in various different arithmetical calculations are very important in arithmetic and mathematics in general. Therefore, Dowker (2005) includes the arithmetic facts among the following three knowledge types that are considered relevant to arithmetic: *factual knowledge, procedural knowledge* and *conceptual knowledge*. This section, according to the findings of recent research, will describe the mechanisms by which number facts are stored in memory. The implications from the dysfunction of these mechanisms for students with learning difficulties will also be presented.

As was previously stated, number facts are stored in long-term memory, but also in semantic or conceptual memory. Individual differences in recalling arithmetic facts from memory are closely related to individual differences in attitudes generally towards arithmetic (Gray & Mulhern, 1995). Pupils with learning difficulties in mathematics struggle with storing and retrieving arithmetical facts from memory and often rely on arithmetical strategies, for example, counting at ages where their peers use more fact retrieval from memory in calculation. Also, for these students, unlike regular students, arithmetic growth does not imply a change from one stage, where problem solving is based on procedural methods, to another stage, where problem solving is based on methods of memory retrieval (Fei, 2000; Geary & Brown, 1991a, 1991b; Geary, Widaman, Little & Cormier, 1987; Ostad, 1997; Russell & Ginsburg, 1984; Yeo, 2001). The majority of the population facing these difficulties may recover some arithmetic facts, but might struggle to retrieve facts associated with an operation (e.g., multiplication). Nevertheless, some children might have a normal capacity to recall facts associated with another operation (i.e., subtraction); this occurs most commonly in cases where the retrieval deficits derive from brain injury (Pesenti, Seron & Van der Linden, 1994).

Equally, it is important to note that fact retrieval features appear more often in some arithmetical situations than others. For instance, the majority of exercises rely on the memorisation of multiplication facts (multiplication tables) more than addition facts. Similarly, more people typically memorise addition facts than subtraction and division facts. This is because for most people, the multiplication depends more on knowledge of facts, while subtraction and further division depend on the procedural knowledge (Dowker, 2005, p. 182).

Geary (2004, 1993) and Geary, Hamson and Hoard (2000) found that students with learning difficulties in mathematics, when retrieving arithmetical facts from long-term memory, make more errors than their peers with normal performance. Furthermore, students with learning difficulties in mathematics demonstrate different patterns of errors and response time compared to younger children with typical performance. However, the patterns of response time for children with learning difficulties are similar to children who have suffered injury to the left hemisphere or related areas of the sub–cortex. Based on this, Geary (2004) concluded that the deficiencies in the memory for many of these children referred to the same mechanisms that create retrieval deficits in dyscalculia. Geary (2004) stated that although the cognitive and neural mechanisms that create these deficits are not completely understood, they are involved in the representation of information of the language system. He bases his assumption in the cognitive mechanisms involved in the formation of representations of arithmetic facts in long-term memory.

In the solution of arithmetic problems using counting procedures, associations are created between problems and resultant responses (Siegler, 1996). Counting typically employs phonological and semantic representational systems of the linguistic domain (e.g., understanding quantities associated with number words). Any rupture in the ability of representation or to retrieve information from these systems would theoretically result in difficulties forming problem–response associations during counting (Geary, 1993, Geary, Bow-Thomas, Fan & Siegler, 1993). This seems to be a case with difficulties in learning arithmetic facts and retrieving those facts that are represented in long-term memory. As a matter of fact, Dehaene et al. (1996) came to similar conclusions. They suggested that the retrieval of arithmetic facts was actually supported by a system of neural structures, which in turn supported the phonological and semantic representations that were bound during growth processes (e.g., counting) (Dehaene & Cohen, 1995, 1997). However, it has not been scientifically confirmed whether the retrieval deficit of children with learning difficulties in mathematics results from damage or neuro-developmental disorders in the areas identified by Dehaene and Cohen (1995, 1997).

Barrouillet, Fayol and Lathulibre (1997) first discovered another form of deficit in children with learning difficulties in mathematics. A deficit in the retrieval of arithmetic facts due to difficulties in inhibition of inappropriate responses. The retrieval of arithmetic facts depends on the presence of knowledge of arithmetic facts in long-term memory. It depends, however, on the ability to choose appropriate numerical facts and the inhibition of alternative facts, which is considered a function of long-term memory (Barrouillet et al., 1997). This memory function is important because when many similar arithmetic facts are stored together, there can easily be confusion between them. Barrouillet et al. (1997) conducted a survey among students, 13 and 14 years old, with learning difficulties in mathematics. He found that they do not face as many difficulties with the knowledge of multiplication facts as with the inhibition of related but incorrect responses. For instance, in the case of multiplication, multiplication facts can be confused with other facts belonging to the same table, especially the ones that immediately precede or follow (e.g., when the answer $6 \times 3 = 24$ is found, there is confusion between the 6×3 and 6×4). This confusion is called associative interference. Similar errors were also stated by Geary et al. (2000) to exist in simple addition problems, for example, in the problem $6 + 2$, a common retrieval error was confusing the numbers 6 and 2 with the numbers 7 and 3 accordingly. As it can be clearly seen these numbers are the next in the sequence. Therefore, such errors are related to the associations in the sequence of numbers (counting-string associations).

7.5. The use of strategies in mental calculations of students with mathematical disabilities

This section will examine the characteristics of mental calculation strategies and their development in students who face difficulties in mathematics.

It is known in the literature that there are some common features among all students with mathematical disabilities (MD). (Geary, 1993; Geary et al., 2000;

Geary, Hoard, Byrd-Craven & DeSoto, 2004; Jordan, Hanich & Kaplan, 2003; Landerl, 2004; Rousselle & Noel, 2008). Specifically, students with MD have a certain difficulty in performing arithmetic processes and learning or retrieving arithmetic facts from memory. The literature suggests that these difficulties can lead to the predominant employment of immature strategies in performing mental calculations.

Recent studies on mental calculation strategies for students with MD discuss simple additions and subtractions with numbers up to 20. In order to present the results of these researches as well as follow the differences in the strategies presented during the development of the learners, the model of Siegler (1995) on strategic change ASCM (Chapter 1, Section 1.4.3) should be recalled. On the basis of this specific model, there are four dimensions, distinguished as follows:

1. the strategic repertoire (which strategies are used);
2. the relative frequency (referring to when each strategy is used);
3. the efficiency of strategy (how each strategy is executed: this refers to the speed and accuracy of implementation of each strategy);
4. the choice of strategy (how strategies are chosen: this refers to the flexibility and adaptivity of the strategy choices of the individual).

Several studies were performed on the characteristics of the strategies students with MD use in the first and second grades in addition and subtraction with numbers up to 20 (Geary, 1990; Geary, Bow-Thomas & Yao, 1992; Geary, Hoard & Hamson, 1999; Hanich, Jordan, Kaplan & Dick, 2001; Jordan & Hanich, 2000; Torbeyns, Verschaffel & Ghesquière, 2004). The results of this research show that students of this age use the same strategy repertoire as students without MD. This repertoire of strategies includes strategies of perceptualisation of numbers, counting strategies, along with retrieval strategies (see Chapter 3).

However, first- and second-grade students with MD use more often immature counting strategies, such as counting with fingers or objects, and less frequently retrieval strategies, compared to students without MD. Furthermore, students with MD make more errors, when using these strategies, than their peers without MD. This dimension relates to the efficiency of the strategy. Also, first- and second-grade students with MD seem to have less flexibility in their choice of strategy than their peers without MD. Students without MD use counting strategies to deal with difficult problems and retrieval strategies to deal with simpler problems, while students with MD use counting strategies to do more problems.

In older pupils – from third to sixth grade – the difference in efficiency and adaptivity of strategies among students with and without MD seems to disappear (Geary & Brown, 1991a, 1991b; Jordan & Montani, 1997; Ostad, 1998). However, while students from third to sixth grade without MD use retrieval strategies in the simplest additions and subtractions, their peers with MD mainly use counting strategies. This indicates that a difference in the relative frequency of strategies, as was explained above, still remains.

Regarding the development of strategies deployed to deal with simple addition and subtraction by students with MD (Geary, Brown & Samaranayake, 1991; Geary et al., 2000; Ostad, 1997), it can be observed that research findings pertaining to the development of strategies seem to be similar to strategy characteristics. Both younger and older elementary school students with MD and without MD apply the same repertoire of strategies, consisting of strategies that vary from counting to retrieval strategies.

However, students without MD, as they age and acquire more experience, use retrieval strategies more often and use counting strategies less. Conversely, students with MD, despite experience gained with time and practising simple additions and subtractions up to 20, presented no changes in the frequency and accuracy of strategies. Students with MD use more counting strategies and fewer retrieval strategies for the majority of problems. However, for students with MD, as they age, they increase their accuracy in counting strategies, while the accuracy in retrieval strategies still remains low. According to the above, Geary (1993, 1994) concluded that students with MD present a delay in the development of strategies in relation to students without MD. This could mean that students with MD develop counting strategies more slowly than their peers without MD. Although the retrieval strategies of students with MD do not grow more slowly compared to students without MD, their strategies develop in a qualitatively different way.

In a recent investigation, Torbeyns et al. (2004) used the choice/no choice method, combined with chronological age (CA), and ability level (AL), to investigate the characteristics of the strategies and their development in students with and without MD to simple additions and subtractions. The results of this research showed that children with MD demonstrated a delay in the development of calculation skills. It was observed that they applied the same strategies with the same frequency, speed, accuracy and adaptivity as younger children corresponding to their mathematical level.

7.6. Teaching interventions for children with difficulties in mathematics, especially in mental calculations

As can be seen above, most typical features of students with learning difficulties in mathematics are associated with skills that are required in mental calculations. It has been observed that students with learning difficulties in mathematics face great difficulty both in storing as well as in retrieving basic arithmetic facts from memory. Generally, students with learning difficulties in mathematics seem to struggle with the use of various strategies and operations that require extensive use of working memory, which renders them prone to making multiple errors and presenting a certain time delay compared to their peers.

Following this, it can be assumed that mental calculations will be a weak point for students with learning difficulties in mathematics, and will therefore require special handling in teaching. In particular, the weaknesses of children with learning difficulties in mathematics do not only affect their behaviour in mental calculations, but also affect their behaviour in arithmetic and generally in mathematics. Therefore, instructional interventions and the enhancement of the respective mathematical contents are deemed essential for the general mathematical formation of these children. Specifically, the formation of early arithmetical concepts constitutes the foundation on which the subsequent mathematical structure is built. Moreover, negative emotions, such as anxiety and low self-esteem, generated by low performance, may cause a generally negative attitude towards learning mathematics.

Modern mathematics curricula, which have recently come to be adopted in many countries, considerably value the learning features and capabilities of each student, unlike previous traditional programmes, which focused only on the mathematical content. In modern programmes, there is an attempt to present mathematics in accordance with the needs of everyday life, taking into account the level and prior knowledge of children. Based on this, it is proposed to implement a teaching process for students in order to construct knowledge as well as interact and participate in a class. Furthermore, education is necessary to include all students, regardless of ethnicity, language, gender, cognitive abilities and any other specificities.

This proposed teaching environment is obviously more favourable and appropriate for students with learning difficulties than that of traditional teaching. Processes of knowledge construction, such as working in groups, communication between students and differentiated and individualised instruction, when properly applied, seem to increase learning opportunities for students with learning difficulties. However, there are studies showing that educational materials and methods recommended in these modern programmes are not completely suitable for children with learning difficulties. For instance, it has been found that programme materials and teaching methodologies based on the NCTM American standards are very complex, unstructured and confusing for many students with different learning needs (Baxter, Woodward & Olson, 2001; Woodward, 2006).

The question, then, whether constructivist approaches and realistic mathematics education (RME, faculty of Holland) are as beneficial for students with poor performance as for students who perform well in mathematics, remains. Van Zoelen, Houtveen and Booij (1997) concluded that, although students with good or average performance benefit from the programme of realistic mathematics of Holland (RME), students with low performance seem to benefit much less from this method. Woodward and Baxter (1997) indicated that special teachers have raised objections to the teaching methods and materials of the NCTM standards programme in the USA because they are too focused on discovery and show no sensitivity to teaching students with disabilities. In their research, the previously mentioned authors also state that this kind of teaching favours most students, but those with learning difficulties and low performance are encouraged to a much lesser extent. Several authors, such as Geary (1994) and Rivera (1997), argue that students with learning difficulties in mathematics require special instruction.

As discussed above, students with difficulties in mathematics significantly vary in needs and performance. On this basis, it can be assumed that a single type of teaching could not satisfy the diversified needs of these students. However, based on various studies on instructional interventions and research summaries, some general recommendations can be made to guide teaching and curricula for students with learning difficulties or difficulties in mathematics.

There are many studies investigating suitable instruction in mathematics for students with learning difficulties. In this issue there are five meta-analyses, which examined 183 research studies (Adams & Carnine, 2003; Baker, Gersten & Lee, 2002; Browder, Spooner, Ahlgrim-Delzell, Harris & Wakeman, 2008; Kroesbergen & Van Luit, 2003; Xin & Jitendra, 1999). According to these studies, the following four methods of teaching are described as the most effective:

- **Systematic and explicit instruction.** Explicit teaching, often called *direct instruction*, refers to a teaching practice that carefully constructs interactions between students and their teacher. Teachers follow a specified educational series to a specific teaching objective. They assess how much students already know about a subject and adjust their teaching accordingly. Systematic teaching focuses on teaching students *how to learn* by providing them with tools and techniques that can be used by students to understand and learn new concepts or skills. Systematic teaching is sometimes called 'strategy instruction' when it refers to the students' learning strategies. These strategies help students integrate new information with what they already know in a meaningful way that enables them later to recall the information or ability, even in a different situation.
- **Self-instruction.** Self-instruction refers to a variety of self-regulation strategies that students can use to manage themselves as learners and to guide their behaviour, including their attention (Graham, Harris & Reid, 1992). With self-instruction, they develop a detailed picture of themselves as learners (metacognitive awareness) and self-regulation skills used by good learners to manage and bear the cost of the learning process.
- **Peer tutoring.** *Peer tutoring* is a term used to describe a wide range of teaching, but most studies include students working in pairs and helping each other to learn and practise concepts that constitute academic objectives. Peer tutoring works best when students with different levels of abilities work together (Kunsch, Jitendra & Sood, 2007). In peer tutoring instruction, it is common for a teacher to switch roles with students, so that teachers can become learners. To put it differently, when someone explains a concept to another person, the persons who explains is equally benefitted in further developing their learning. This method helps students to have a better understanding of the content they are studying.
- **Visual representations.** The final form of mathematical concepts is abstract; in fact, the more complicated mathematic concepts get, the more abstract they are. Most students find abstract concepts difficult to manage. Based on this, visual representations can assist learning, especially when mathematics is only presented in an abstract form. Under a broad definition, visual representations may include manipulatives, pictures, diagrams, charts and the number line,

among others. Among the most common examples of mathematics instruction that involve visual representations are Concrete–Representational–Abstract (CRA) techniques. CRA techniques have proven to be very effective teaching methods for mathematics for students with disabilities (Butler, Miller, Crehan, Babbitt & Pierce, 2003; Morin & Miller, 1998).

The CRA is a three-part instructional strategy. The first step involves using *concrete* materials (such as the abacus, Unifix cubes, geometric figures, geoboards) to model mathematical concepts that are being taught; for example, the number 26 is formed by Unifix codes. In the second part, the concepts are presented in representational form (such as images and shapes), for example, 26 is represented with a picture showing 26 items organised into tens and units. Finally, the concepts are presented in an abstract or symbolic form (such as numbers and mathematical symbols). In this case, 26 are represented by digits (26) or a number word (twenty-six).

7.6.1. Studies on teaching interventions in mental calculations

Montague (2011), noted that the most effective interventions for students with learning disabilities were direct instruction and cognitive strategy instruction. Direct instruction is based on behavioural theory, while cognitive strategy instruction includes behavioural and cognitive theory. Direct instruction is more didactic than cognitive strategy instruction and related more to the instruction of basic skills (Kroesbergen & Van Luit, 2003). It uses scripted lessons, which are guided by the teacher and fast-paced. In contrast, cognitive strategy instruction is interactive and uses explicit teaching, which focuses on teaching students the processes involved in the application of skills, such as solving mathematical problems.

7.6.1.1. Studies of direct teaching

Studies that use direct instruction mainly focus on drill and practise to improve the retrieval of arithmetic facts and computational skills. Drills and practises are used to improve the recall and automaticity of arithmetic facts by students with learning disabilities (LD) because they provides challenge, appropriate time on task and numerous response opportunities.

The beneficial effects of drill–rehearsal methods led to the development of specific approaches to practice. Two such examples are the 'Drill Sandwich' (Coulter & Coulter, 1990) and the 'Incremental Rehearsal' (Tucker, 1989). The 'Drill Sandwich' (Coulter & Coulter, 1990) comprised 50% known and 50% unknown situations, while the 'Incremental Rehearsal' (Tucker, 1989), used a gradually increasing ratio of known to unknown situations reaching the final stage of implementing the proportion of 90% to 10%. By comparing the Drill Sandwich, the Incremental Rehearsal and traditional practice (100% unknown), it was found that Incremental Rehearsal leads to significantly better retention in memory than the other two models after 1, 2, 3, 7 and 30 days (MacQuarrie et al., 2002).

A typical example of this model of drill–rehearsal is the incremental rehearsal used by Burns (2005). Burns (2005) studied teaching instructions of multiplication tables for third-grade students identified as learning disabled in mathematical computation. In the incremental rehearsal model, as mentioned above, a proportional increase gradually from known to unknown facts is used until students reach at least 90% success. Ten new multiplication facts are introduced each time. In this study, Incremental Rehearsal appeared to be an effective intervention for increasing fluency of single-digit multiplication facts among these third-grade children.

Another study of direct instruction investigated the capability of second-grade students with and without learning disabilities to add single-digit numbers (Tournaki, 2003). In this study, a specific teaching strategy for addition was compared to the method of drill and practise. In detail, students with learning disabilities were taught to use the strategy of counting from the largest to minimum number. Initially, students identified the larger addend number and then counted upwards as many steps as the smallest number; for example, when adding 3 + 5, students started from 5 and went up three steps: (5), 6, 7, 8. Results showed that students who were taught the strategy of counting from the largest number improved significantly compared to those who learned basic arithmetic facts by the method of drill and practise.

7.6.1.2. Research in cognitive strategy instruction

There are some studies that use a more cognitive approach to the instruction of disabled students. Instruction in self-regulation is an important technique of cognitive strategy. Van Luit, Kaskens and van der Krol (1993), recognising the importance of self-regulation methods in increasing the use of strategies in mathematics, developed a special training programme to teach multiplication and division for children with difficulties in mathematics with an emphasis on the use of strategy. This programme was called MASTER (Mathematics Strategy Training for Educational Remediation) and was based on the assumption that strategy instruction in mathematics could help special children improve their behaviour in mathematics.

Van Luit and Naglieri (1999) used the programme MASTER, which relied on self-regulation, to improve the skills of elementary school third-grade students in basic multiplication and division. This study investigated the effect of the MASTER programme on 84 students, of whom 42 had learning difficulties and the remaining 42 mild mental retardation. The MASTER programme included an introductory phase, a phase of training in a group and an individual practise phase. Finally, it followed a discussion between teacher and students of all possible solution procedures. Students were taught to identify different strategies and then decide which was the most appropriate for the solution of specific problems. The survey results showed that the use of the self-instruction programme resulted in significant improvement over the general instruction programme. Moreover, far transfer was found for the children with learning disabilities in the experimental group when they used effective problem-solving strategies on non-trained tasks.

In another study, the MASTER programme was used for teaching multiplication to 75 pupils aged 7–13 years, 27 of whom faced learning difficulties (Kroesbergen & Van Luit, 2002). Students were taught by the MASTER programme, by direct instruction and by the regular curriculum, the Dutch Realistic Mathematics Education (RME) programme. Regarding the MASTER programme, self-instruction was promoted by means of a strategy decision sheet, which contained several questions that the students were posed, such as 'What is the multiplication problem?', 'Do I know the answer directly?', 'Do I know the answer if I reverse the problem?', 'Do I know the answer if I say the multiplication table aloud?', and 'Do I know the answer if I do long addition?' Later the page was extended and included the following strategies: splitting by five or ten, using a neighbouring problem, and doubling. The strategy decision sheet assisted students to learn to use various strategies. Students taught to use the MASTER programme outnumbered those who had received only regular curriculum training and had better behaviour in transferring the gained knowledge compared to the two groups of students taught with direct instruction or who followed the regular curriculum. The results showed that students without learning disabilities benefited more from the MASTER programme than through direct instruction in learning basic multiplications and divisions, while students with learning disabilities benefited more from direct instruction.

7.6.2. Mnemonic strategies

As was previously mentioned, failure of memory constituted one of the key characteristics of people with learning disabilities. In the literature of special education there are a number of investigations, referring to mnemonic strategies, which are ways of instruction that improve student performance (e.g., Mastropieri & Scruggs, 2000; Scruggs & Mastropieri, 2000).

Mnemonic strategy is an instructional strategy commonly used by students with and without special needs to improve memory. Mnemonic strategies are systematic procedures for enhancing memory by providing effective cues, such as words, phrases or pictures. Mnemonic strategies are usually separated into virtual representations, such as pictures or diagrams, and verbal compositions, using words to help memory (Scruggs & Mastropieri, 1990).

The significance of mnemonic strategies in school life and their use by children with learning disabilities began to be investigated and commonly established from the early '80s. However, the existence and use of these strategies were set even further in the past. The first documented use of mnemonic strategies was among ancient Greeks. In ancient times, having little access to materials and tools to assist writing, humans developed a complex mnemonic system for the storage and recall from memory of stories, poems and lectures. A common strategy used to remember information was the 'method of loci' (from the Latin word *locus*, which means space). This method was first used by the poet Simonides, who was able to recognise the mangled corpses through the ruins of a Roman palace, because he remembered the place where everyone was sitting before the disaster. So in the context of the loci method, space is used to connect objects with specific places. (See Yates 1966 for more information on the history of mnemonic techniques.)

Nowadays, there are at least three distinct methods for teaching mnemonic strategies: *The keyword method, the peg-word method* and *letter strategies*. These methods are briefly presented below.

- **The keyword method:** The keyword method is probably the most versatile of the mnemonic strategies and is useful when students need to remember new vocabulary or names of new facts and concepts (Scruggs & Mastropieri, 1990). This method is based on the association of new information with keywords that already exist in memory. To teach a new word that is difficult for students, first a similar sounding keyword has to be identified and then it has to be represented by a picture or drawing. Hence, students create a mental image that connects the word to be learnt with its definition. When students need to recall the original word from memory, they can imagine the image associated with the keyword, for example, for someone to remember the Italian word *rosso* (= red), they can imagine a Russian with red clothes.
- **The peg-word method:** The peg-word method uses a correspondence/rhyme between numbers and words (one is *bun*, two is *shoe*, three is *tree*, etc.), so that students remember information numbered or arranged in series. For instance, to help a student remember that insects have six legs, an image presenting insects on tree branches (= sticks, peg-word for 6) can be used. Also, a picture of a spider in a gate (= gate, peg-word for 8) can help students to remember that spiders have eight legs.
- **Letter strategies:** Letter strategies include the use of *acronyms*. Acronyms are words the letters of which represent elements in lists of information. Kilpatrick (1985) noted the use of the acronym FOIL, to recall the process of distributive properties in multiplying two binomials (a + b) (c + d). According to the acronym for calculation of this expression initially we multiply the first terms (**F**irst terms) ac, then the two external (**O**uter terms) ad, then the two inner (**I**nner terms) bc and the last terms (**L**ast terms) bd. Finally, we add the products to find the result of the algebraic expression. Another example of an acronym (Miller & Mercer, 1993) is **DRAW,** which describes a problem-solving technique.

    Discover the sign.
    Read the problem.
    Answer or draw and check.
    Write the answer.

- **Yodai methods.** The Yodai method is a Japanese system, linking abstract concepts to known pictures or stories. This method seems to lead to a better understanding of complex operations. To illustrate, Higbee and Kunihira (1985) taught fraction operations to young students as follows: the fraction was termed an 'insect', the head of which was named numerator while as denominator were the wings. Adding fractions with a common denominator was expressed by 'counting heads when the wings are the same'. Thus, the connection of abstract ideas with known images or stories led to a better understanding of complex operations (Higbee & Kunihira, 1985).

Scruggs and Mastropieri (1990, p. 271), defined a mnemonic as 'a device, procedure or operation that is used to improve memory'. Mnemonics can be distinguished into two types: *fact mnemonics* and *process mnemonics*. Fact mnemonics are the most common form used for memorisation of arithmetic facts, namely to construct a mnemonic association for each item to be remembered. In contrast, mnemonic processes are less known formats used to remember rules, principles and procedures (Manalo, Bunnell & Stillman, 2000, p. 138).

In Japan, a type of mnemonic processes is the method Yodai, mentioned above. This method has successfully been used for many years in teaching various subjects, such as arithmetic, algebra, geometry, trigonometry, calculus, organic and inorganic chemistry, physics, biology, spelling, grammar as well as the English language. Manalo et al. (2000) used an adapted mnemonic process from Japanese to teach the four operations to thirteen- and fourteen-year-old students with learning disabilities in mathematics. In this instruction, numbers were presented as characters (warriors) and operations as stories. For instance, in subtraction, the top number represents the attackers and the lower number the defenders warriors. Conclusively, the results showed that the method of process mnemonics helps students with learning disabilities to improve.

7.6.2.1. The use of mnemonic strategies in teaching multiplication tables

In several studies, mnemonic strategies were used to improve the memory retrieval capacity of multiplication facts. These studies focused primarily on students with learning disabilities (Greene, 1992, 1999; Irish, 2002; Wood, Frank & Wacker, 1998; Wood & Frank, 2000; Zisimopoulos, 2010).

Greene (1999) and Wood et al. (1998) used the peg-word method for the instruction of multiplication tables with students with learning disabilities. This method will be presented below as it was described by Wood and Frank (2000):

Families of basic multiplication facts. The 100 multiplication facts involving the numerals 0–9, from 0×0 up to 9×9, are grouped into six families: the zeros, ones, twos, fives, nines, and pegwords (the 15 remaining facts). In order for a fact to belong in one of the above families, it has to have one of its two digits associated with the name of the family. For example, 0×4, belongs to the family of 0, while 1×9 can belong both in the family 1 and 9.

Strategies to remember the families of multiplication tables. The facts containing 0 and 1 are easier and therefore prior to learning compared to the families of 2, 5 and 9. Peg-words were taught last, because they were the most difficult family. Students practise the strategies of families of prep using the strategy cards.

For example, in the card of strategy of 0 dictates: *Look for a 0 (top or bottom). If you find it . . . the answer is 0.* The strategy for the family of peg-words is more complicated and therefore its teaching requires more time. There are cards where, on one side, the number is presented with digits as well as number-words while on the other side a picture-symbol which rhymes is presented. For example (one–bun, two–shoe,..., twelve–elf, . . . , thirty–dirty, . . .). Additionally, the results of the products are presented on cards with corresponding pictures, for example the fact $4 \times 3 = 12$ is shown with a picture of a tree (three–tree) with a door (four–door) where an elf enters (twelve–elf).

Irish (2002) used a multimedia software program to teach students with learning disabilities and middle mental retardation how to use the mnemonic strategies of keywords and peg-words to learn multiplication facts. After this intervention, although students succeeded in correctly retrieving a large number of multiplication facts, students with middle mental retardation failed to achieve a score higher than 85%.

In his research Zisimopoulos (2010) used an instruction comprising a variant of mnemonic peg-word strategy in combination with a picture fading technique. He specifically examined the effectiveness of this instruction with two students with moderate mental challenges in the retrieval of 28 multiplication facts between 2 and 9. The results of this investigation showed that both students improved their accuracy in the retrieval of multiplication facts, while they equally maintained the new knowledge and were able to generalise into other areas.

References

Adams, G. & Carnine, D. (2003). Direct instruction. In H. L. Swanson, K. R. Harris & S. Graham (Eds.), *Handbook of learning disabilities* (pp. 403–416). New York: Guilford Press.

American Psychiatric Association (1994). *Diagnostic and statistical manual of mental disorders* (4th ed.). Washington, DC: American Psychiatric Association.

Atkinson, R. C. & Shiffrin, R. M. (1968). Human memory: A proposed system and its control processes. In K. W. Spence & J. T. Spence (Eds.), *The psychology of learning and motivation* (Vol. 2., pp. 89–195). New York: Academic Press.

Baddeley, A. D. (1986). *Working memory*. Oxford: Oxford University Press.

Baddeley, A. D. (1990). *Human memory: Theory and practice*. Hove, U.K.: Erlbaum.

Baddeley, A. D. (1996). Exploring the central executive. *Quarterly Journal of Experimental Psychology, 49A*(1), 5–28.

Baddeley, A. D. (2000). The episode buffer: A new component of working memory. *Trends in Cognitive Sciences, 4*, 417–423.

Baddeley, A. D. & Hitch, G. J. (1974). Working memory. In G. H. Bower (Ed.), *The psychology of learning and motivation: Advances in research and theory* (Vol. 8, pp. 47–90). New York: Academic Press.

Badian, N. A. (1983). Arithmetic and nonverbal learning. In H. R. Myklebust (Ed.), *Progress in learning disabilities* (pp. 253–264). New York: Grune & Stratton.

Baker, S., Gersten, R. & Lee, D. (2002). A synthesis of empirical research on teaching mathematics to low-achieving students. *The Elementary School Journal, 103*(1), 51–73.

Barrouillet, P., Fayol, M. & Lathulibre, E. (1997). Selecting between competitors in multiplication tasks: An explanation of the errors produced by adolescents with learning disabilities. *International Journal of Behavioral Development, 21*, 253–275.

Baxter, J., Woodward, J. & Olson, D. (2001). Effects of reform-based mathematics instruction in five third-grade classrooms. *Elementary School Journal, 101*, 529–548.

British Dyslexia Association. Definitions. Retrieved from http://www.bdadyslexia.org.uk/dyslexic/definitions.

Browder, D. M., Spooner, F., Ahlgrim-Delzell, L., Harris, A. & Wakeman, S. Y. (2008). A meta-analysis on teaching mathematics to students with significant cognitive disabilities. *Exceptional Children, 74*(4), 407–432.

Burns, M. K. (2005). Using incremental rehearsal to increase fluency of single-digit multiplication facts with children identified as learning disabled in mathematics computation. *Education and Treatment of Children, 28*, 237–249.

Butler, F. M., Miller, S. P., Crehan, K., Babbitt, B. & Pierce, T. (2003). Fraction instruction for students with mathematics disabilities: Comparing two teaching sequences. *Learning Disabilities Research & Practice, 18*(2), 99–111.

Butterworth, B., Grana, A., Piazza, M., Girelli, L., Price, C. & Skuse, D. (1999). Language and the origins of numbers skills: Karyotypic differences in Turner's syndrome. *Brain and Language, 69*, 486–488.

Bynner, J. and Parsons, S. (2005). *Does numeracy matter more?* London: Institute of Education, National Research and Development Centre for Adult Literacy and Numeracy.

Chinn, S. J. & Ashcroft, J. R. (1993). *Mathematics for dyslexics*. London: Whurr.

Cohen, L., Dehaene, S., Chochon, F., Lehricy, S. & Naccache, L. (2000). Language and calculation within the parietal lobe: A combined cognitive, anatomical and fMRI study, *Neuropsychologia, 38*, 1426–1440.

Coulter, W. A. & Coulter, E. M. (1990). *Curriculum-based assessment for instructional design*. Unpublished manuscript.

Craick, F. I. & Lockhart, R. S. (1972). Levels of processing: A framework for memory research. *Journal of Verbal Learning and Verbal Behavior, 11(6)*, 671–684.

Dehaene, S. & Cohen, L. (1995). Towards an anatomical and functional model of number processing. *Mathematical Cognition, 1*, 83–120.

Dehaene, S. & Cohen, L. (1997). Cerebral pathways for calculation: Double dissociation between rote verbal and quantitative knowledge of arithmetic. *Cortex, 33*, 219–250.

Dehaene, S., Molko, N., Cohen, L. & Wilson, A. J. (2004). Arithmetic and the brain. *Current Opinion in Neurobiology, 14*(2), 218–224.

Dehaene, S., Piazza, M., Pinel, P. & Cohen, L. (2003). Three parietal circuits for number processing. *Cognitive Neuropsychology, 20*(3–6), 487–506.

Dehaene, S., Spelke, E., Pinel, P., Stanescu, R. & Tsivkin, S. (1999). Sources of mathematical thinking: Behavioral and brain imaging evidence. *Science, 284*, 970–974.

Dehaene, S., Tzourio, N., Frank, V., Raynaud, L., Cohen, L., Mehler, J. & Mazoyer, B. (1996). Cerebral activations during number multiplication and comparison: A PET study. *Neuropsychologia, 34*, 1097–1116.

Delazer, M., Domahs, F., Bartha, L., Brenneis, C., Lochy, A., Trieb, T. & Benke, T. (2003). Learning complex arithmetic – An fMRI study. *Cognitive Brain Research, 18*(1), 76–88.

Delazer, M., Ischebeck, A., Domahs, F., Zamarian, L., Koppelstaetter, F., Siedentopf, C. M., Kaufmann, L., Benke, T. & Felber, S. (2005). Learning by strategies and learning by drill – evidence from an fMRI study. *NeuroImage, 25*(3), 838–849.

DfES (2001). *Guidance to support pupils with dyslexia and dyscalculia*. London: Department of Education and Skills.

Dowker, A. (2005). *Individual differences in arithmetic: Implications for psychology, neuroscience, and education*. Hove, NY: Psychology Press.

Fei, H. F. (2000). Why they fall behind in math: A study of underlying cognitive factors of second-grade high-math achievers and low-math achievers in Taiwan. *Dissertation Abstracts International, 60*, 43–56.

Geary, D. C. (1990). A componential analysis of an early learning deficit in mathematics. *Journal of Experimental Child Psychology, 49*, 363–383.

Geary, D. C. (1993). Mathematical disabilities: Cognitive, neuropsychological, and genetic components. *Psychological Bulletin, 114*, 345–362.

Geary, D. C. (1994). *Children's mathematical development: Research and practical applications.* Washington, DC: American Psychological Association.

Geary, D. C. (2000). Mathematical disorders: An overview for educators. *Perspectives, 26*(3), 6–9.

Geary, D. C. (2003). Math disabilities. In H. L. Swanson, K. R. Harris & S. Graham (Eds.), *Handbook of learning disabilities* (pp. 199–212). New York: Guilford.

Geary, D. C. (2004). Mathematics and learning disabilities. *Journal of Learning Disabilities, 37*(1), 4–15.

Geary, D. C. & Brown, S. C. (1991a). Cognitive addition: Strategy choices and speed-of-processing differences in gifted, normal and mathematically disabled children. *Developmental Psychology, 27*, 398–406.

Geary, D. C. & Brown, S. C. (1991b). Strategy choices and speed-of-processing differences in gifted, normal and mathematically disabled children. *Developmental Psychology, 27*, 787–797.

Geary, D. C., Brown, S. C. & Samaranayake, V. A. (1991). Cognitive addition: A short longitudinal study of strategy choice and speed-of-processing difference in normal and mathematically disabled children. *Developmental Psychology, 27*, 787–797.

Geary, D. C., Bow-Thomas, C. C., Fan, L. & Siegler, R. S. (1993). Even before formal instruction, Chinese children outperform American children in mental addition. *Cognitive Development, 8*, 517–529.

Geary, D. C., Bow-Thomas, C. C. & Yao, Y. (1992). Counting knowledge and skill in cognitive addition: A comparison of normal and mathematically disabled children. *Journal of Experimental Child Psychology, 54*, 372–391.

Geary, D. C., Hamson, C. O. & Hoard, M. K. (2000). Numerical and arithmetical cognition: A longitudinal study of process and concept deficits in children with learning disability. *Journal of Experimental Child Psychology, 77*, 236–263.

Geary, D. C., Hoard, M. K., Byrd-Craven, J. & DeSoto, M. C. (2004). Strategy choices in simple and complex addition: Contributions of working memory and counting knowledge for children with mathematical disability. *Journal of Experimental Child Psychology, 88*, 121–151.

Geary, D. C., Hoard, M. K. & Hamson, C. O. (1999). Numerical and arithmetical cognition: Patterns of functions and deficits in children at risk for a mathematical disability. *Journal of Experimental Child Psychology, 74*, 213–239.

Geary, D. C., Widaman, K. F, Little, T. D. & Cormier, P. (1987). Cognitive addition: Comparison of learning disabled and academically normal elementary school children. *Cognitive Development, 2*, 249–269.

Gilbert, A. M. (1992). *A status study of dyscalculia for the primary grades* (unpublished doctoral dissertation). Temple University, Philadelphia, Penn.

Grabner, R. H., Ansari, D., Reishofer, G., Stern, E., Ebner, F. & Neuper, C. (2007). Individual differences in mathematical competence predict parietal brain activation during mental calculation. *NeuroImage, 38*(2), 346–356.

Graham, S., Harris, K. R. & Reid, R. (1992). Developing self-regulated learners. *Focus on Exceptional Children, 24*(6), 1–16.

Gray, C. & Mulhern, G. (1995). Does children's memory for addition facts predict general mathematical ability? *Perceptual and Motor Skills, 81*, 163–167.

Greene, G. (1992). Multiplication facts: Memorization made easy. *Intervention in School and Clinic, 27*, 150–154.

Greene, G. (1999). Mnemonic multiplication fact instruction for students with learning disabilities. *Learning Disabilities Research & Practice, 14*, 141–148.

Gross, J. (2009). *The long term costs of numeracy difficulties*. London: Every Child Chance Trust (KPMG).

Hanich, L. B., Jordan, N. C., Kaplan, D. & Dick, J. (2001). Performance across different areas of mathematical cognition in children with learning difficulties. *Journal of Educational Psychology, 93*, 615–626.

Henschen, S. E. (1925). Clinical and anatomical contributions on brain pathology. *Archives of Neurology and Psychiatry, 13*, 226–249.

Higbee, K. L. & Kunihira, S. (1985). Cross-cultural applications of yodai mnemonics in education. *Educational Psychologist, 20*, 57–64.

Irish, C. (2002). Using peg- and keyword mnemonics and computer-assisted instruction to enhance basic multiplication performance in elementary students with learning and cognitive disabilities. *Journal of Special Education Technology, 17*(4), 29–40.

Ischebeck, A., Zamarian, L., Egger, K., Schocke, M. & Delazer, M. (2007). Imaging early practice effects in arithmetic. *NeuroImage, 36*(3), 993–1003.

Ischebeck, A., Zamarian, L., Siedentopf, C., Koppelstatter, F., Benke, T., Felber, S. & Delazer, M. (2006). How specifically do we learn? Imaging the learning of multiplication and subtraction. *NeuroImage, 30*(4), 1365–1375.

Jordan, N. C. & Hanich, L. B. (2000). Mathematical thinking in second-grade children with different forms of LD. *Journal of Learning Disabilities, 33*, 567–578.

Jordan, N. C., Hanich, L. B. & Kaplan, D. (2003). Arithmetic fact mastery in young children: A longitudinal investigation. *Journal of Experimental Child Psychology, 85*, 103–119.

Jordan, N. C. & Montani, T. O. (1997). Cognitive arithmetic and problem solving: A comparison of children with specific and general mathematics difficulties. *Journal of Learning Disabilities, 30*, 624–634, 684.

Kazui, H., Kitagaki, H. & Mori, E. (2000). Cortical activation during retrieval of arithmetical facts and actual calculation: A functional magnetic resonance imaging study. *Psychiatry and Clinical Neuroscience, 54*, 479–485.

Kilpatrick, J. (1985). Doing mathematics without understanding it: A commentary on Higbee and Kunihira. *Educational Psychologist, 20*(2), 65–68.

Kosc, L. (1970). Psychology and psychopathology of mathematical abilities. *Studia Psychologica, 12*, 159–162.

Kosc, L. (1974). Developmental Dyscalculia. *Journal of Learning Disabilities, 7*, 164–177.

Kroesbergen, E. H. & Van Luit, J. E. H. (2002). Teaching multiplication to low-math performers: Guided versus structured instruction. *Instructional Science, 30*, 361–378.

Kroesbergen, E. H. & Van Luit, J. E. H. (2003). Mathematics interventions for children with special educational needs. *Remedial and Special Education, 24*(2), 97–114.

Kunsch, C., Jitendra, A. & Sood, S. (2007). The effects of peer mediated instruction in mathematics for students with learning problems: A research synthesis. *Learning Disabilities Research & Practice, 22*(1), 1–12.

Lefevre, J. A., Smith-Chant, B., Hiscock, K., Daley, K. & Morris, J. (2003). Young adults' strategies in simple arithmetic: Implications for the development of mathematical representation. In A. Baroody & A. Dowker (Eds.), *The development of arithmetical concepts and skills* (pp. 203–228). Mahweh, NJ: Lawrence Erlbaum Associates, Inc.

MacQuarrie, L. L., Tucker, J. A., Burns, M. K. & Hartman, B. (2002). Comparison of retention rates using traditional, Drill Sandwich, and Incremental Rehearsal flashcard methods. *School Psychology Review, 31*, 584–595.

Manalo, E., Bunnell, J. K. & Stillman, J. A. (2000). The use of process mnemonics in teaching students with mathematics learning disabilities. *Learning Disability Quarterly, 23*(2), 137–156.

Mastropieri, M. A. & Scruggs, T. E. (2000). *Teaching students ways to remember: Strategies for learning mnemonically*. Cambridge, MA: Brookline Books.

Mazzocco, M. (2001). Math learning disability and math LD subtypes: Evidence from studies of Turner syndrome, Fragile X syndrome, and neurofibromatosis type 1. *Journal of Learning Disabilities*, *34*, 520–533.

Menon, V., Rivera, S. M., White, C. D., Eliez, S., Glover, G. H. & Reiss, A. L. (2000). Functional optimization of arithmetic processing in perfect performers. *Cognitive Brain Research*, *9*(3), 343–345.

Menon, V., Rivera, S. M., White, C. D., Glover, G. H. & Reiss, A. L. (2000). Dissociating prefrontal and parietal cortex activation during arithmetic processing. *NeuroImage*, *12*(4), 357–365.

Miles, T. R. (1992). Some theoretical considerations. In T. R. Miles & E. Miles (Eds.), *Dyslexia and Mathematics* (pp. 1–18). London: Routledge.

Miller, S. P. & Mercer, C. D. (1993). Mnemonics: Enhancing the math performance of students with learning difficulties. *Intervention in School and Clinic*, *29*(2), 78–82.

Miyake, A., Friedman, N. P., Emerson, M. J., Witzki, A. H., Howerter, A. & Wager, T. D. (2000). The unity and diversity of executive functions and their contributions to complex frontal lobe tasks: A latent variable analysis. *Cognitive Psychology*, *41*, 49–100.

Montague, M. (2011). Effective instruction in mathematics for students with learning difficulties. In C. Wyatt-Smith, J. Elkins & S. Gunn (Eds.), *Multiple perspectives on difficulties in learning literacy and numeracy* (pp. 295–313). Dordrecht, the Netherlands: Springer.

Morin, V. A. & Miller, S. P. (1998). Teaching multiplication to middle school students with mental retardation. *Education and Treatment of Children*, *21*, 22–36.

Ostad, S. A. (1997). Developmental differences in addition strategies: A comparison of mathematically disabled and mathematically normal children. *British Journal of Educational Psychology*, *67*, 345–357.

Ostad, S. A. (1998). Developmental differences in solving simple arithmetic word problems and simple number-fact problems: A comparison of mathematically normal and mathematically disabled children. *Mathematical Cognition*, *4*, 1–19.

Parkin, A. (1993). *Memory: Phenomena, experiment and theory*. Oxford: Blackwell.

Pesenti, M., Seron, X. & Van der Linden, M. (1994). Selective impairment as evidence for mental organization of arithmetical facts: BB, a case of preserved subtraction? *Cortex*, *30*, 661–671.

Price, G. R. & Ansari, D., (2013). Dyscalculia: Characteristics, causes, and treatments. *Numeracy*, *6*(1), 1–16.

Rivera, D. P. (1997). Mathematics education and students with learning disabilities: Introduction to the special series. *Journal of Learning Disabilities*, *30*, 2–19.

Roland, P. E. & Friberg, L. (1985). Localization of cortical areas activated by thinking. *Journal of Neurophysiology*, *53*(5), 1219–1243.

Rousselle, L. & Noël, M-P. (2008). Mental arithmetic in children with mathematics learning disabilities: The adaptive use of approximate calculation in an addition verification task. *Journal of Learning Disabilities*, *41*(6), 498–513.

Rueckert, L., Lange, N., Partiot, A., Appollonio, I., Litvan, I., Le Bihan, D. & Grafman, J. (1996). Visualizing cortical activation during mental calculation with functional MRI. *NeuroImage*, *3*(2), 97–103.

Russell, R. & Ginsburg, H. P. (1984). Cognitive analysis of children's mathematical difficulties. *Cognition and Instruction*, *1*, 217–244.

Shalev, R. S., Manor, O., Kerem, B., Ayali, M., Badichi, N., Friedlander, Y. & Gross-Tsur, V. (2001). Developmental dyscalculia is a familial learning disability. *Journal of Learning Disabilities*, *34*, 59–65.

Siegler, R. S. (1995). How does change occur: A micro genetic study of number conservation. *Cognitive Psychology, 28,* 225–273.
Siegler, R. S. (1996). *Emerging minds: The process of change in children's thinking.* New York: Oxford University Press.
Siegler, R. S. & Jenkins, E. (1989). *How children discover new strategies.* Hillsdale, NJ: Lawrence Erlbaum Associates, Inc.
Scruggs, T. & Mastropieri, M. (1990). Mnemonic instruction for students with learning disabilities: What it is and what it does. *Learning Disability Quarterly, 13*(4), 271–281.
Scruggs, T. & Mastropieri, M. (2000). The effectiveness of mnemonic instruction for students with learning and behavior problems: An update and research synthesis. *Journal of Behavioral Education, 10*(2/3), 163–173.
Torbeyns, J., Verschaffel, L. & Ghesquière, P. (2004). Strategy development in children with mathematical disabilities: Insights from the choice/no-choice method and the chronological-age/ability-level-match design. *Journal of Learning Disabilities, 37*(2), 119–31.
Tournaki, N. (2003). The differential effects of teaching addition through strategy instruction versus drill and practice to students with and without learning disabilities. *Journal of Learning Disabilities, 36*(5), 449–458.
Tucker, J. A. (1989). *Basic flashcard technique when vocabulary is the goal* (unpublished teaching materials). University of Tennessee at Chattanooga.
Van Luit, J. E. H., Kaskens, J. & Van der Krol, R. (1993). *Special rekenhulpprogramma vermenigvuldigen en verdelen.* [MASTER strategy training programme.] Doetinchem, The Netherlands: Graviant.
Van Luit, J. E. H. & Naglieri, J. A. (1999). Effectiveness of the MASTER program for teaching special children multiplication and division. *Journal of learning disabilities, 32*(2), 98–107.
Van Zoelen, E. M., Houtveen, A. A. M. & Booij, N. (1997). *Evaluatie project kwaliteitsversterking rekenen en wiskunde: Het eerste projectjaar.* [Evaluation of the Project Quality Reinforcement in Mathematics: The First Year.] Utrecht, The Netherlands: Utrecht University/ISOR.
Venkatraman, V., Siong, S. C., Chee, M. W. L. & Ansari, D. (2006). Effect of language switching on arithmetic: A bilingual fMRI study. *Journal of Cognitive Neuroscience, 18*(1), 64–74.
Xin, Y. P. & Jitendra, A. K. (1999). The effect of instruction in solving mathematical word problems for students with learning problems: A meta-analysis. *The Journal of Special Education, 32*(4), 207–225.
Wood, D. & Frank, A. (2000). Using memory – enhancing strategies to learn multiplication facts. *Teaching Exceptional Children, 32.*
Wood, D., Frank, A. & Wacker, D. (1998). Teaching multiplication facts to students with learning disabilities. *Journal of Applied Behavior Analysis, 31,* 323–338.
Woodward, J. (2006). Making reform-based mathematics work for academically low-achieving middle school students. In M. Montague & A. Jitendra (Eds.), *Teaching mathematics to middle school students with learning difficulties* (pp. 29–50). New York: Guilford Press.
Woodward, J. & Baxter, J. (1997). The effects of an innovative approach to mathematics on academically low-achieving students in inclusive settings. *Exceptional Children, 63,* 373–388.
World Health Organization (1994). *International Classification of Diseases* (10th ed.). Geneva: WHO.
Yates, F. (1966). *The art of memory.* Chicago: The University of Chicago Press.
Yeo, D. (2001). *Dyslexia and mathematics.* Paper presented at the 5th British Dyslexia Association Conference, London.

Zago, L., Pesenti, M., Mellet, E., Crivello, F., Mazoyer, B. & TzourioMazoyer, N. (2001). Neural correlates of simple and complex mental calculation. *NeuroImage, 13* (2), 314–327.

Zago, L. & Tzourio-Mazoyer, N., (2002). Distinguishing visuospatial working memory and complex mental calculation areas within the parietal lobes. *Neuroscience Letters, 331*(1), 45–49.

Zisimopoulos, D. (2010). Enhancing multiplication performance in students with moderate intellectual disabilities using pegword mnemonics paired with a picture fading technique. *Journal of Behavioral Education, 19(2)*, 117–133.

INDEX

abacus 80–81
acalculia 209
accuracy 16, 29, 31–34; addition and subtraction 82; estimation 202; learning difficulties 218; rational numbers 173
acquired dyscalculia 209
acronyms 224
action schemas 129
Adaptive Strategy Choice Model (ASCM) 33, 34
adaptivity 28, 29, 32, 33; adaptive expertise 29, 49–53, 98; estimation 190; learning difficulties 217, 218
adding up strategy 65, 94
addition 3, 73–114; adaptive expertise 50; brain studies 214; bus model 60; column calculation 66; empty number line 60–65; estimation 188, 189, 191–192, 193, 195, 199, 201, 203; Greek curriculum 20–21; inherent skills 19; learning difficulties 217–218, 222; memorisation 215; Nature and Life Mathematics research 85–88, 99–110; rational numbers 161, 168, 171, 177; sets and cardinality 20; strategies 24, 26, 29, 32, 34, 46–48, 73–98, 120, 121, 132; subtraction by addition 91, 94, 96–98; written algorithms 26; *see also* repeated addition

additive doubling 118, 120, 121, 131
adjusting strategy 189
age 21, 22, 190–191, 193, 202, 218
Alajmi, A. H. 195, 196
algebra 9, 161–162, 225
algorithms 1–2, 8, 11, 17, 24, 25–28, 29, 45; addition and subtraction 82; estimation 189, 194, 197; fractional division 161; instrumental understanding 14; mental image of 91, 95; multiplication 133, 139–140, 142, 143–146, 147–153; Nature and Life Mathematics research 101–106, 107, 108; Netherlands 55; rational numbers 169, 170, 176–177; teaching 46
American Psychiatric Association 207–208
Anghileri, J. 7, 14, 117, 119
angular gyrus 214
Appollonio, I. 214
approximation 185
Archimedes 148, 149
arithmetic facts: brain processes 214; estimation 190; Identical Elements model 131; memory 214–216; mnemonics 225; multi-digit multiplication and division 136; multiplication tables 116, 117, 118–119, 123, 124, 125
articulatory loop 212
Ashcraft, M. H. 85

Index

Ashcroft, J. R. 210
associations model 85
associativity 17, 190
Atkinson, R. C. 210–211
attitude 82, 219
Australia 10, 49, 133, 168, 170
averaging 188, 199, 200, 201
Ayali, M. 213

Baddeley, A. D. 211, 212
Badichi, N. 213
Baek, J.-M. 133
Bajic, D. 132–133
Ball, D. L. 169
Barger, R. H. 17
Barrouillet, P. 216
Baxter, J. 219
Behr, M. 169
Beishuizen, M. 45, 46, 88, 90, 94
Belgium 98, 192, 193
benchmark strategy 165, 166, 170, 174, 176, 189
Berch, D. 14
Blöte, A. W. 30, 45, 46, 50, 94
Bobis, J. 203
Booij, N. 219
Booth, J. L. 185, 186, 190
Booth, L. R. 162
brain studies 213–214
Brazil 36
bridging through ten strategy 30, 36n2, 50, 78, 79, 80–81, 90–91, 93
Brown, G. W. 2, 6
Brown, M. 163
Brown, S. 129
Bryant, P. 163, 170
Burns, M. K. 222
bus model 60
Butterworth, B. 19–20, 22
Buys, K. 64

calculation, definition of 8
calculators 1, 2, 6, 10, 29
Callingham, R. 167, 168
Campbell, J. I. D. 131
Caney, A. 164, 167
cardinality 19, 20, 22, 76, 82–83
Carpenter, T. P. 47, 83
Carr, M. 96
Carraher, D. W. 17

Carraher, T. N. 17
Case, R. 13
Castro, E. 195
central executive system 211–212
change 11, 21
child development 20–22
Children's Math Worlds (CMW) 119
China 35, 192–193
Chinn, S. J. 210
choice/no choice method 33, 108–109, 218
chronometric models 85
circles 149
Clarke, D. M. 170
classroom context 45
clustering 188, 198, 199, 201
Cobb, P. 53, 88
Cockroft Report (1982) 11
cognitive abilities 9
cognitive development 31, 34
cognitive psychology 19, 34
cognitive strategy instruction 221, 222–223
Cohen, L. 85, 216
collections 90
column calculation 66–67, 147–153
common content knowledge 169
communities of practice 48, 49
commutativity 17, 125, 130, 190
compatible numbers strategy 189, 195, 199, 200, 201
compensation: addition 24, 78, 79, 90, 91, 94; estimation 187, 188–189, 193, 198, 199; multi-digit multiplication and division 133, 135, 138, 142, 143–146; Realistic Mathematics Education 65; subtraction 30
computation, definition of 8
computational estimation 4, 7, 9, 183–206; definition of 8; instructional interventions 202–204; modern international programs 10; Nature and Life Mathematics research 196–202; number sense 17–18; quantitative reasoning 12; research on students' abilities 194–195; strategies 34, 187–189, 193–195, 198–202, 203; teachers 195–196, 199–202; variables 189–194
computational resources 118, 123
computational strategies 116
conceptual change theories 162

conceptual knowledge 215
conceptual strategies 15–16, 164, 169; *see also* number-sense strategies
conceptual understanding 14–16, 84, 164
Concrete-Representational-Abstract (CRA) techniques 221
conservation 22
construction strategies 74, 78–80, 84
constructivism 2, 48, 53, 219
context 45, 48, 192
context variables 29, 34–36, 190, 192–194
Cooney, J. B. 121
Cooper, R. G. 21
Cooper, T. J. 15, 46
count strategies 24, 25, 31; addition and subtraction 73–74, 76–78, 82–84, 87, 89–91, 95; counting all strategy 24, 73, 74, 75, 76, 82, 83–84; counting on from larger (min) strategy 24, 31, 76, 83–84, 85, 117; counting on from the first (max) strategy 31, 74, 82, 83; learning difficulties 217, 218; multi-digit multiplication and division 133, 134, 136; multiplication tables 118, 119, 120–121, 124; simple division 130
counting 19–20, 21–22; cultural context 35; empty number line 61, 62–63; learning difficulties 216, 222; multiplication tables 117, 118; National Numeracy Strategy 56, 58–59; Nature and Life Mathematics research 101–106; Realistic Mathematics Education 57, 59, 60
counting all strategy 24, 73, 74, 75, 76, 82, 83–84
counting on from larger (min) strategy 24, 31, 76, 83–84, 85, 117
counting on from the first (max) strategy 31, 74, 82, 83
covert strategies 96
Craick, F. I. 211
cultural context 34–35, 36
curricula 2, 110, 183, 219

data handling 9
De Brauwer, J. 123, 132
De Corte, E. 14
De Lange, J. 12
De Nardi, E. 96, 164
De Smedt, B. 94, 96, 97–98

Deboys, M. 88
decimals 17, 23, 159, 160, 168; errors 2; estimation 191, 195, 199; front-end strategy 188; Greek curriculum 20–21; Nature and Life Mathematics research 171–174; proportional reasoning 164; strategies 167; student difficulties with 163; *see also* rational numbers
decomposition 17, 18, 123
Dehaene, S. 85, 214, 216
Deri, M. 116
derived fact strategies 74, 78–80, 84, 130
derived products 120, 121, 122, 141–142
developmental dyscalculia 209, 213
developmental theories 162
'didactic contract' 176
direct counting 118, 130
direct instruction 220, 221–222, 223
direct modelling 130, 133, 134, 136, 139
direct retrieval: addition and subtraction 74, 84, 88, 89; multi-digit multiplication and division 134, 136, 141–142; multiplication tables 120, 121–122; simple division 132
direct subtraction (DS) 96, 97, 98
disposition 14
distributivity 17, 125, 189, 190, 195
division 4, 22–23, 127–133; estimation 191–192, 195, 203; Greek curriculum 21; initial knowledge of students 129–130; inverse nature of multiplication and 125, 131; learning difficulties 223; Nature and Life Mathematics research 141, 143–146; procedural knowledge 215; rational numbers 161, 168, 177; semantic structure 128; sets and cardinality 20; strategies 7, 24, 130–131, 133–146; written algorithms 67
Dominick, A. 26, 27
doubles 78, 79, 84, 122, 130
doubling: learning difficulties 223; multi-digit multiplication and division 136, 138, 143, 144; multiplication tables 118, 120, 121, 125; simple division 131
Dowker, A. 215
'Drill Sandwich' 221
drill-rehearsal 221–222
Dunphy, E. 13, 14
dyscalculia 20, 208, 209–210, 213, 215
dyslexia 208, 210

Ellis, S. 34–35
Emori, H. 96
England: Cockroft Report 11; estimation 183; flexibility 49; learning difficulties 207; mental calculation in official documents 7; multiplication 153; National Numeracy Strategy 10, 46, 53, 55–60, 66; splitting strategy 90; strategies 45–46; stringing strategy 92; written algorithms 66–67
episodic buffer 212
episodic memory 211
equivalence classes 160
errors 2, 36; estimation 185, 194; learning difficulties 215, 216, 217, 218; multiplication 123, 140; rational numbers 176–177; semantic memory 209; visuo-spatial memory 210; written algorithms 27
estimation 4, 7, 9, 183–206; definition of 8, 185; instructional interventions 202–204; modern international programs 10; Nature and Life Mathematics research 196–202; number sense 17–18; quantitative reasoning 12; research on students' abilities 194–195; strategies 34, 187–189, 193–195, 198–202, 203; teachers 195–196, 199–202; types of 184–185; variables 189–194
Eutocius of Ascalon 148–150, 152
everyday life 9
explicit instruction 220

factorisation 189
factual knowledge 215
family context 35–36
Fayol, M. 216
Fias, W. 123, 132
fingers 45, 56, 73–74, 75, 77–78, 83, 86, 89
Fischbein, E. 116, 117–118, 128
five, using 78, 79, 122, 126
flexibility 9, 24, 28–36; estimation 190, 193, 203; learning difficulties 217; number sense 13, 16, 17; research 108–110; subtraction by addition 97, 98; teaching 46, 49–53, 108
fluency 1, 13, 50
fractions 23, 159, 160–162, 168; definition of 160; errors 2; estimation 189, 191, 195, 200; Greek curriculum 20–21; mental comparison of 169–170; Nature and Life Mathematics research 173–178; strategies 16, 164–166, 170; student difficulties with 162–164; Yodai methods 224–225; *see also* rational numbers
France 121, 193
Frank, A. 225
Freudenthal, H. 54, 59
Friberg, L. 214
Friedlander, Y. 213
front-end strategy 188, 195, 198, 199, 200, 201
Fuchs, L. S. 190
Fuller, D. 96
Fuson, K. C. 82–83, 88, 115, 118, 119–120, 121

Galfano, G. 116
Gallistel, C.-R. 21–22
Geary, D. C. 162, 215, 216, 218, 219
Gelman, R. 19, 21–22, 83, 162
gender 96
geometry 9, 153, 210, 225
Germany 49, 50, 52
Ghesquière, P. 94, 96, 97–98
Gliner, G. S. 192
Goodman, T. 192
Goos, M. 49
Grafman, J. 214
Gray, E. 45
Greece: addition and subtraction 20, 85–88, 99–110; Cross Curriculum Framework 10; estimation 195, 196–202; flexibility 108–110; fractions 163; mnemonic strategies 223; multiplication 122, 153; Nature and Life Mathematics research 1, 85–88, 99–110, 174–175, 196–202; rational numbers 169, 174–175, 177; social factors 35–36; splitting strategy 90; stringing strategy 92
Greek multiplication 147–153
Greene, G. 225
Greer, B. 14, 128
Griva, H. 122
Groen, J. 84, 85
Gross-Tsur, V. 213

Hackling, M. 203–204
halving 125, 130, 131, 136, 138, 143, 144

Hamson, C. O. 215
Hart, K. 163
Hatano, Giyoo 29, 51
Heinze, A. 52
Heirdsfield, A. M. 15, 46, 47–48, 133
Henschen, S. E. 209
Hiebert, J. 15, 48
Higbee, K. L. 224–225
Hitch, G. J. 211
Hoard, M. K. 215
holistic strategies 91, 93–94, 96; flexibility 109–110; multi-digit multiplication and division 133, 135, 138, 142, 143–147; Nature and Life Mathematics research 101–106, 107–108; *see also* compensation
Holyoak, Keith 29
Houtveen, A. A. M.. 219
Hundreds-Tens-Units (HTU) 58–59
Hurry, J. 163
hybrid strategies 119, 120, 121

Identical Elements (IE) model 131
Imbo, I. 192, 193
Inagaki, Kayoko 29
'Incremental Rehearsal' 221–222
indifferent strategies 172–174
indirect addition (IA) 96, 97
indirect subtraction (IS) 91, 94–95, 97
informal knowledge 19
informal mathematics 19
informal strategies 44, 55, 59, 60, 138–140
information processing 211
inhibition 212
innumeracy 10
instrumental strategies 96, 164, 167
instrumental understanding 14
integers 160, 162, 168, 170, 188, 199, 214
intuition 18, 188
intuitive models 115, 116, 117–118, 130
inverse relations 162
investigative approach 52, 53
Irish, C. 226
Irons, C. 46

Japan 96, 183, 224–225
Jenkins, E. 29
Jessup, D. L. 96
'jottings' 7
jumps 62, 63–64, 92

Kaiafa, I. 165, 167, 175
Kaimakami, A. 195
Kalchman, M. 13
Kamii, C. 26, 27, 47
Kaskens, J. 222
Kerem, B. 213
Kermeli, A. 169, 177
Kerslake, D. 163
keyword method 224
Kilpatrick, J. 224
Klein, A. S. 30, 45, 46, 90
knowledge: common and specialised content knowledge 169; division 129–130; estimation 186, 190, 203; informal 19; multiplication 118, 124; strategic 44–45; types of 215
Kosc, L. 209
Kouba, V. L. 119
Kuchermann, D. 163
Kunihira, S. 224–225
Kuwait 195
Kwak, J. 132–133
Kwon, Y. 82–83

Lange, N. 214
language 35, 192, 193
Lathulibre, E. 216
Lave, J. 48
Le Bihan, D. 214
learning 32, 48–49
learning difficulties 4–5, 207–232; brain studies 213–214; definition of terms 207–210; Greek multiplication 153; memory 215–216; number sense 14; strategies 216–218, 223–226; subtraction by addition 98; teaching interventions 218–226
learning-teaching trajectories 55, 56–57
Lecacheur, M. 202
LeFevre, J.-A. 119, 122, 123, 192, 193–194
Lemaire, P. 32–33, 108, 121, 193–194, 202
Lemonidis, Ch. 122, 148, 165, 167, 169, 175, 177, 195, 196, 199
length 184
Lesh, R. 169
letter strategies 224
levelling strategy 65, 78, 79, 91, 94
Levine, D. R. 18, 194
Ligouras, G. 35–36, 108
literacy 10

Litvan, I. 214
Liu, F. 192
Liu, W. 192–193
loci method 223–224
Lockhart, R. S. 211
logic 11
Luwel, K. 190
Lynch, K. 190, 203

Mamede, E. 170
Manalo, E. 225
Manor, O. 213
Marino, M-S. 116
Markovits, Z. 163, 170
MASTER (Mathematics Strategy Training for Educational Remediation) program 222–223
Mastropieri, M. 225
materials 73–76, 80, 86, 89, 221
mathematical learning difficulties *see* learning difficulties
mathematical literacy 10, 11
mathematisation 55, 58, 60, 64
max strategy *see* counting on from the first strategy
McIntosh, A. 2, 12–13, 14, 15, 16, 96, 164
measurement 184, 185
mediation strategy 131
memorisation 84, 123, 124, 126–127, 215
memory 22, 45, 210–213; brain studies 214; division 132; episodic 211; learning difficulties 218, 223; long-term 17, 23–24, 49, 74, 84, 117, 121, 141, 211, 212–213, 215–216; mnemonic strategies 223–226; multiplication 116–117, 121–122, 132; number facts 214–216; procedural 209–210; semantic 209, 211, 215; short-term 7, 9, 24, 80, 84, 211, 212; working 17, 84, 117, 210, 211–213, 218; *see also* retrieval strategies
Menne, J. 62
mental calculation 1–3; conceptual understanding and procedural skills 14–16; definition of 6–8; empty number line 60–66; flexibility 28–36; Greek multiplication 152–153; importance of 8–9; learning difficulties 207–232; modern international programs 9–10; National Numeracy Strategy 55, 56, 59–60; Nature and Life Mathematics research 85–88; number sense 12–14, 16–17; numeracy 10–12; rational numbers 159–182; Realistic Mathematics Education 54–55, 57, 58, 59, 60; strategies 23–25; teacher training 99–108; teaching 44–72; written algorithms compared to 25–28
mental images 45, 91, 95
mental representations 165, 166, 176
metacognition 9, 29, 45, 87, 107, 127
Mickes, L. 132–133
Mildenhall, P. 203–204
Miles, T. R. 210
min strategy *see* counting on from larger strategy
Mitchelmore, M. 117–118, 128, 130
mixed method 91, 93
mnemonic strategies 223–226
Montague, M. 221
Moser, J. M. 83
Moss, J. 13
Mouratoglou, A. 199
Mulligan, J. 117–118, 128, 130
multiplication 3–4, 22–23, 115; column calculation 66–67; estimation 188, 189, 191–192, 195, 201, 202, 203; Greek curriculum 21; Greek multiplication 147–153; holistic strategies 108–110; inverse nature of division and 125, 131; learning difficulties 216, 223, 225–226; memorisation 215; mnemonic strategies 225–226; Nature and Life Mathematics research 138–147; rational numbers 161, 168, 177; semantic structure 128; sets and cardinality 20; shared memory with division 132; strategies 7, 24, 32, 34, 133–147
multiplication tables 3–4, 20, 115–127; brain processes 214; Incremental Rehearsal 222; memory mechanisms 116–117; mnemonic strategies 225–226; strategies 116, 117–123, 124, 125–127, 225–226; teaching 123–127
multiplicative operation 118, 120, 130, 131
Murphy, C. 48–49, 55, 57, 58, 66

Nailon, C. 133
National Curriculum 55
National Numeracy Strategy 10, 46, 53, 55–60, 66

natural numbers 23, 161, 162
Nature and Life Mathematics (NaLiMa) 1; addition and subtraction 85–88, 99–110; estimation 196–202; multi-digit multiplication and division 138–147; rational numbers 171–178
near-doubles 24, 78, 79
Neber, H. 192–193
Nello, M-S. 116
Netherlands 10; division 127–128; flexibility 49; mixed method 93; multiplication 153; Realistic Mathematics Education 46, 53–66, 219, 223; stringing strategy 92; written algorithms 66, 67
nice number strategies 195, 204n1
Nikolantonakis, K. 148, 196
Noelting, G. 164
Nohda, N. 96
Nolka, E. 196
non-numerical estimation 185
Northcote, M. 2
number facts 7, 17, 56, 84–85, 214–216; *see also* arithmetic facts
number facts retrieval model 85
number lines 220; bridging through ten strategy 81; empty 55, 56, 58, 59–66, 81; National Numeracy Strategy 56; number line estimation 185; stringing strategy 92
number sense 2–3, 9, 12–14, 16–17; estimation 202; number-sense strategies 15–16, 164–167, 168–169, 172–178; rational numbers 168–169, 170–171; strategies 15–16; teachers' lack of 169
numeracy 2, 10–12, 13
numerical estimation 185
numerical skills 18–23
numerosity 19–21, 22, 34, 184–185
Nunes, T. 36, 163, 170

objects, using 73–74, 75, 76, 83, 89
openness 192
oral methods 56
ordinality 19, 82
Oura, Y. 51
Overlapping Wave Model 31–32
overt strategies 96

Palaigeorgiou, G. 169, 177
Pan, T. R. 191, 195

Panou, F. 122
paper-and-pencil methods 1, 6, 25
parents 35–36
parietal cortex 213–214
Parkman, J. M. 84, 85
Partiot, A. 214
partitioning number strategies 133, 134–135, 137, 139, 142, 143–147; *see also* splitting strategy
partitive division 128–129, 136
part-whole numbers *see* rational numbers
pattern learning 118
Paulos, J. A. 10
peer tutoring 220
peg-word method 224, 225–226
Peltenburg, M. 98
percentages 23, 159, 160, 168; errors 2; Greek curriculum 21; proportional reasoning 164; strategies 167; *see also* rational numbers
perceptualisation of numbers 74–76, 81–82, 83, 217
Perova, N. 190, 203
Peters, G. 94, 97–98
phonological loop 212
phonological storage 117
Piaget, J. 23
PISA *see* Program for International Student Assessment
Pitt, E. 88
place value 17, 55, 58–59; estimation 183; partitioning number strategies 134, 137, 139, 146, 147
placement of numbers on number line 63
Pnevmatikos, D. 199
Poland 192–193
Post, T. R. 165, 169
precision 192
Pretzlik, U. 163
Primary National Strategy (PNS) 66
privilege theory 162
problem size effect 122, 123, 132
problem solving 9; estimation 183, 187, 196–202; flexibility 30; memory 215; Realistic Mathematics Education 54; strategies 52–53; written algorithms 26
procedural knowledge 215
procedural memory 209–210
procedural skills 14–16, 50–51, 84
procedural strategies *see* rule-based strategies

proceduralisation 55
Program for International Student Assessment (PISA) 11
progressive mathematisation 55, 58, 60, 64
proportion 23
proportional reasoning 159, 164
psychology 19

quantity 11–12, 19, 86–87; estimation of 184–185; extensive and intensive 23
quotitive division 129, 136

Radatz, H. 88
range strategy 189
ratio 22–23
rational numbers 4, 23, 159–182; definition and properties of 160–164; Greek curriculum 21; instructional interventions in number sense 170–171; Nature and Life Mathematics research 171–178; strategies 24, 164–167, 168–169, 170, 172–178; student difficulties with 162–164
ratios 164
Realistic Mathematics Education (RME) 46, 53–66, 219, 223
reformulation 187, 189
relational strategies 96, 164
relational understanding 14
relationships 11
repeated addition 22–23; multi-digit multiplication and division 139, 141–142, 143–146; multiplication tables 118, 120, 121, 122, 123, 124; simple division 130, 131
repeated subtraction 130, 131, 141–142, 143–146
repetition 52
residual thinking 165, 166, 170, 176
Resnick, L. B. 18, 84
retrieval models 115, 116
retrieval strategies: addition and subtraction 32, 74, 83, 84, 85, 88, 89, 96; learning difficulties 217, 218; multi-digit multiplication and division 136, 141–142; multiplication tables 120, 121–122; simple division 132
revised Identical Elements (IE-r) model 131, 132
Reys, B. J. 17, 96, 202

Reys, R. E. 7, 8–9, 17, 96, 170
rhythmic counting 117, 118, 120, 121, 122, 130–131
Rickard, T. C. 131, 132–133
Rittle-Johnson, B. 190, 203
Rivera, S. M. 219
Robinson, K. 132
Roche, A. 170
Roland, P. E. 214
rounding 188, 193–194, 195, 198, 199, 200–202, 203
routine approach 52
routine expertise 29, 49–50
Ruddock, G. 163
Rueckert, L. 214
rule-based strategies 15–16, 164–167, 169, 172–178, 194

Saldanha, L. A. 164
Schipper, W. 88
Schliemann, A. D. 17
Schoenfeld, A. H. 184–185
School Curriculum and Assessment Authority (SCAA) 46
Scruggs, T. 225
Seethaler, P. M. 190
self-efficacy 191, 192
self-instruction 220, 223
self-regulation 222
Selter, C. 94
semantic memory 209, 211, 215
semantic representations 86–87
semantic types 115
semiotic representations 161
sequencing method *see* stringing strategy
sets 20, 22
Sfard, A. 49
Shalev, R. S. 213
shapes 11
sharing 128–129, 136
Sherin, B. 115, 118, 119–120, 121
Shiffrin, R. M. 210–211
shifting 212
Shrager, J. 85
Siegler, R. S. 29, 31–34, 85, 108, 121, 123, 185, 186, 190, 217
Skemp, R. R. 14, 15, 96, 164
skills 18–23
skip counting 130, 131
social goals 35

sociocultural perspective 34, 53
solution procedures 115, 116
Sowder, J. T. 8, 18, 163, 169–170, 184, 190, 191, 194
space 11
special numbers strategy 189, 195, 198, 199, 200, 201
specialised content knowledge 169
specific disorder of mathematical skills 208
speed 29, 31–34; addition and subtraction 82; estimation 191, 202; holistic strategies 109, 110; learning difficulties 218
split-jump method 93
splitting (1010) strategy 26, 88–92, 96; column calculation 66; England 57, 58–59; flexibility 30; learning difficulties 223; Nature and Life Mathematics research 101–106, 107, 108; problem-solving approach 53; Realistic Mathematics Education 46, 57, 59, 65
Stafylidou, S. 163
Star, J. R. 190, 203
Starkey, P. 21, 83
Stassens, N. 94, 96, 97
strategies 7, 17, 23–25, 26, 28; addition and subtraction 73–98; choice of 32–34, 217; classification of 95–96, 119–120; cognitive strategy instruction 221, 222–223; comparison of England and Netherlands 56–60; cultural context 34–35, 36; division 130–131, 132, 133–146; estimation 18, 187–189, 193–195, 198–202, 203; family context 35–36; flexibility 28, 30, 50–51, 52–53, 108–110; informal 44, 55, 59, 60, 138–140; learning difficulties 216–218, 223; mnemonic strategies 223–226; multi-digit multiplication 133–147; multiplication tables 116, 117–123, 124, 125–127, 225–226; Nature and Life Mathematics research 85–88, 101–106; Overlapping Wave Model 31–32; procedural or conceptual 15–16; rational numbers 164–167, 168–169, 170, 172–178; Realistic Mathematics Education 54–55; systematic instruction 220; teaching 44–48, 83, 110
Strategy Choice and Discovery Simulation (SCADS) 31

stringing (N10) strategy 24, 26, 88–90, 91, 92–93, 96; England 57, 58–59; flexibility 30, 50; Nature and Life Mathematics research 101–106; problem-solving approach 53; Realistic Mathematics Education 46–47, 57, 59, 65
subject variables 29, 31–34, 190–191
substitution 189
subtraction 3, 73–114; adaptive expertise 50; brain studies 214; bus model 60; column calculation 66; empty number line 60–61; estimation 191–192, 203; Greek curriculum 20–21; indirect 91, 94–95, 97; inherent skills 19; learning difficulties 217–218; mnemonic strategies 225; Nature and Life Mathematics research 85–88, 99–110; procedural knowledge 215; rational numbers 161, 168, 171, 177; repeated 130, 131; sets and cardinality 20; strategies 24, 29, 30, 32, 34, 46–48, 73–98; subtraction by addition 91, 94, 96–98; written algorithms 26–27
subtraction by addition (SA) 91, 94, 96–98
sum strategy see counting all strategy
Swan, P. 96, 164
systematic teaching 220

Taiwan 168–169, 171, 194, 195
task variables 29, 30, 190, 191–192
teacher training 99–108, 169, 203
teaching 44–72; comparison of England and Netherlands 53–67; conceptual understanding 15; estimation 202–204; flexibility and adaptive expertise 49–53, 108; international research 44–53; learning difficulties 218–226; multiplication tables 123–127; Nature and Life Mathematics research 87; strategies 44–48, 83, 110; written algorithms 27
technology 10
ten, multiplying by 125–126
Ter Heege, Hans 124, 126
textbooks 54, 55, 153
Thompson, Ian 7, 9, 25, 66, 88
Thompson, P. W. 164
Threadgill-Sowder, J. 18
Threlfall, J. 25, 51
ties effect 122

Torbeyns, J. 94, 96, 97–98, 190, 218
Trachilou, E. 138
Trafton, P. R. 6
transitive property 165
translation 187
Treffers, A. 61, 90
truncation 188, 193
Tsakiridou, H. 122
Tsao, Y. L. 191, 195, 196
Tussendoelen Annex Leerlijnen (TAL) 56

uncertainty 11
unitary counting 117, 118, 130
United States: addition strategies 83; China compared with 35; estimation 183; flexibility 49; NCTM standards 10, 219; rational numbers 159–160
updating 212

Van den Berg, E. 30
Van den Heuvel-Panhuizen, M. 54, 96, 190, 192, 203
Van der Heijden, M. K. 30
Van der Krol, R. 222
Van Dooren, W. 190
Van Luit, J. E. H. 222
Van Mulken, F. 90
Van Putten, K. 90
Van Zoelen, E. M. 219
varying strategy 57, 96
Vergnaud, G. 128
Verguts, T. 123
Verschaffel, L. 14, 28–30, 49, 50–53, 94, 96–98, 190
visual representations 59, 220–221
visuo-spatial memory 210

visuo-spatial sketchpad 212
Vosniadou, S. 163

Wachsmuth, I. 169
Walker, D. 132–133
Wandt, E. 2, 6
Watson, J. M. 167, 168
Wearne, D. 15, 48
Wheeler, M. M. 190, 191, 194
Whitacre, I. 16, 171
whole number analogues 167
Williams, E. 162
Wood, D. 225
Woods, S. S. 84
Woodward, J. 219
word problems 47, 48, 115, 128
working memory 17, 84, 117, 210, 211–213, 218
World Health Organization (WHO) 208
written algorithms 1–2, 8, 17, 24, 25–28, 29, 45; addition and subtraction 82; division 127–128; England 66–67; errors 36; mental image of 91, 95; multiplication 143–146, 147–153; Nature and Life Mathematics research 101–106, 107, 108; Netherlands 55, 66, 67; rational numbers 165, 169, 170, 176–177; teaching 46
Wu, S. S. 192
Wynn, K. 21

Yackel, E. 90
Yang, D. C. 164, 168–169, 171, 192, 194
Yodai methods 224–225

Zisimopoulos, D. 226